Z

Physics and
Fractal Structures

Physics and Fractal Structures

Jean-François GOUYET

Condensed matter physics, École polytechnique (Palaiseau, France)

Foreword by Benoît MANDELBROT

IBM Watson Research Center, Yale University

Including 4 color plates

 Springer

New York Berlin Heidelberg Barcelona Budapest Hong Kong
London Milan Paris Santa Clara Singapore Tokyo

Paris Milan Barcelone

Published with the support of ministère de l'Enseignement supérieur et de la Recherche (France) :
Direction de l'information scientifique et technique et des bibliothèques.

Cover illustration: Computer-enhanced synthetic-aperture radar image of a region in the St. Elias
Mountains in Alaska and Canada. Colors represent radar backscatter intensity. The image covers
an area roughly 100 km across and shows the Bagley ice field (vertical blue region at right), and
the Logan and Walsh glaciers. The image was acquired 28 May 1992 by the European Space
Agency's First Remote Sensing Satellite, ERS-1. Processing by K. Ahlns was supported by a
NASA Polar Research Program grant to C.S. Lingle and W.D. Harrison, Geophysical Institute,
University of Alaska. SAR image © ESA, 1992. Reproduced with permission.

ISBN 3-540-94153-3 Springer-Verlag Berlin Heidelberg New York
ISBN 0-387-94153-3 Springer-Verlag New York Berlin Heidelberg
ISBN 2-225-85130-1 Masson

Printed in France

MASSON
SPRINGER-VERLAG

120, bd Saint-Germain, 75280 Paris Cedex 06
Heidelberger Platz 3, D-1000 Berlin 33
Tiergartenstrasse 17, D-6900 Heidelberg 1

Foreword

When intellectual and political movements ponder their roots, no event looms larger than the first congress. The first meeting on fractals was held in July 1982 in Courchevel, in the French Alps, through the initiative of Herbert Budd and with the support of IBM Europe Institute. Jean-François Gouyet's book reminds me of Courchevel, because it was there that I made the acquaintance and sealed the friendship of one of the participants, Bernard Sapoval, and it was from there that the fractal bug was taken to École Polytechnique. Sapoval, Gouyet and Michel Rosso soon undertook the work that made their laboratory an internationally recognized center for fractal research. If I am recounting all this, it is to underline that Gouyet is not merely the author of a new textbook, but an active player on a world-famous stage. While the tone is straightforward, as befits a textbook, he speaks with authority and deserves to be heard.

The topic of fractal diffusion fronts which brought great renown to Gouyet and his colleagues at Polytechnique is hard to classify, so numerous and varied are the fields to which it applies. I find this feature to be particularly attractive. The discovery of fractal diffusion fronts can indeed be said to concern the theory of welding, where it found its original motivation. But it can also be said to concern the physics of (poorly) condensed matter. Finally it also concerns one of the most fundamental concepts of mathematics, namely, diffusion. Ever since the time of Fourier and then of Bachelier (1900) and Wiener (1922), the study of diffusion keeps moving forward, yet entirely new questions come about rarely. Diffusion fronts brought in something entirely new.

Returning to the book itself, if the variety of the topics comes as a surprise to the reader, and if the brevity of some of treatments leaves him or her hungry for more, then the author will have achieved the goal he set himself. The most important specialized texts treating the subject are carefully referenced and should satisfy most needs.

To sum up, I congratulate Jean-François warmly and wish his book the great success it deserves.

Benoît B. MANDELBROT

Yale University
IBM T.J. Watson Research Center

...

So, Nat'ralists observe, a Flea
Hath smaller Fleas that on him prey,
And these have smaller yet to bite 'em
And so proceed ad infinitum.

...

Jonathan Swift, 1733,
On poetry, a Rhapsody.

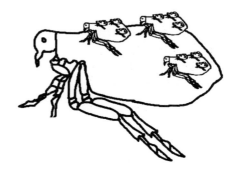

Contents

Foreword v

Preface xiii

1. Fractal geometries

1.1 Introduction 1
1.2 The notion of dimension 2
1.3 Metric properties: Hausdorff dimension, topological dimension 4
    1.3.1 The topological dimension 5
    1.3.2 The Hausdorff – Besicovitch dimension 5
    1.3.3 The Bouligand – Minkowski dimension 6
    1.3.4 The packing dimension 8
1.4 Examples of fractals 10
    1.4.1 Deterministic fractals 10
    1.4.2 Random fractals 19
    1.4.3 Scale invariance 21
    1.4.4 Ambiguities in practical measurements 22
1.5 Connectivity properties 23
    1.5.1 Spreading dimension, dimension of connectivity 23
    1.5.2 The ramification $\mathcal{R}$ 25
    1.5.3 The lacunarity $\mathcal{L}$ 25
1.6 Multifractal measures 26
    1.6.1 Binomial fractal measure 27
    1.6.2 Multinomial fractal measure 30
    1.6.3 Two scale Cantor sets 36
    1.6.4 Multifractal measure on a set of points 38

2. Natural fractal structures: From the macroscopic...

2.1 Distribution of galaxies 41
    2.1.1 Distribution of clusters in the universe 42
    2.1.2 Olbers' blazing sky paradox 43
2.2 Mountain reliefs, clouds, fractures... 45
    2.2.1 Brownian motion, its fractal dimension 46
    2.2.2 Scalar Brownian motion 48
    2.2.3 Brownian function of a point 49
    2.2.4 Fractional Brownian motion 49
    2.2.5 Self-affine fractals 52
    2.2.6 Mountainous reliefs 57
    2.2.7 Spectral density of a fractional Brownian motion, the spectral
    exponent β 58
    2.2.8 Clouds 61
    2.2.9 Fractures 62
2.3 Turbulence and chaos 65
    2.3.1 Fractal models of developed turbulence 66
    2.3.2 Deterministic chaos in dissipative systems 72

3. Natural fractal structures:... to the microscopic

3.1 Disordered media 89
   3.1.1 A model: percolation 89
   3.1.2 Evaporated films 105
3.2 Porous media 107
   3.2.1 Monophasic flow in poorly connected media 108
   3.2.2 Displacement of a fluid by another in a porous medium 109
   3.2.3 Quasistatic drainage 111
3.3 Diffusion fronts and invasion fronts 118
   3.3.1 Diffusion fronts of noninteracting particles 118
   3.3.2 The attractive interaction case 125
3.4 Aggregates 130
   3.4.1 Definition of aggregation 130
   3.4.2 Aerosols and colloids 132
   3.4.3 Macroscopic aggregation 140
   3.4.4 Layers deposited by sputtering 141
   3.4.5 Aggregation in a weak field 142
3.5 Polymers and membranes 146
   3.5.1 Fractal properties of polymers 146
   3.5.2 Fractal properties of membranes 152

4. Growth models

4.1 The Eden model 157
   4.1.1 Growth of the Eden cluster: scaling laws 159
   4.1.2 The Williams and Bjerknes model 163
   4.1.3 Growing percolation clusters 164
4.2 The Witten and Sander model 165
   4.2.1 Description of the DLA model 165
   4.2.2 Extensions of the Witten and Sander model 167
   4.2.3 The harmonic measure and multifractality 172
4.3 Modeling rough surfaces 174
   4.3.1 Self-affine description of rough surfaces 174
   4.3.2 Deposition models 174
   4.3.3 Analytical approach to the growth of rough surfaces 176
4.4 Cluster – cluster aggregation 177
   4.4.1 Diffusion – limited cluster – cluster aggregation 177
   4.4.2 Reaction – limited cluster – cluster aggregation 179
   4.4.3 Ballistic cluster – cluster aggregation and other models 180

5. Dynamical aspects

5.1 Phonons and fractons 183
   5.1.1 Spectral dimension 183
   5.1.2 Diffusion and random walks 188
   5.1.3 Distinct sites visited by diffusion 191
   5.1.4 Phonons and fractons in real systems 192
5.2 Transport and dielectric properties 194
   5.2.1 Conduction through a fractal 194
   5.2.2 Conduction in disordered media 197
   5.2.3 Dielectric behavior of composite media 207
   5.2.4 Response of viscoelastic systems 208

5.3 Exchanges at interfaces 211
    5.3.1 The diffusion-limited regime 213
    5.3.2 Response to a blocking electrode 213
5.4 Reaction kinetics in fractal media 215

6. Bibliography 219

7. Index 231

Preface

The introduction of the concept of fractals by Benoît B. Mandelbrot at the beginning of the 1970's represented a major revolution in various areas of physics. The problems posed by phenomena involving fractal structures may be very difficult, but the formulation and geometric understanding of these objects has been simplified considerably. This no doubt explains the immense success of this concept in dealing with all phenomena in which a semblance of disorder appears.

Fractal structures were discovered by mathematicians over a century ago and have been used as subtle examples of continuous but *nonrectifiable* curves, that is, those whose length cannot be measured, or of continuous but *nowhere differentiable* curves, that is, those for which it is impossible to draw a tangent at any their points. Benoît Mandelbrot was the first to realize that many shapes in nature exhibit a fractal structure, from clouds, trees, mountains, certain plants, rivers and coastlines to the distribution of the craters on the moon. The existence of such structures in nature stems from the presence of disorder, or results from a functional optimization. Indeed, this is how trees and lungs maximise their surface/volume ratios.

This volume, which derives from a course given for the last three years at the Ecole Supérieure d'Electricité, should be seen as an introduction to the numerous phenomena giving rise to fractal structures. It is intended for students and for all those wishing to initiate themselves into this fascinating field where apparently disordered forms become geometry. It should also be useful to researchers, physicists, and chemists, who are not yet experts in this field.

This book does not claim to be an exhaustive study of all the latest research in the field, yet it does contains all the material necessary to allow the reader to tackle it. Deeper studies may be found not only in Mandelbrot's books (Springer Verlag will publish a selection of books which bring together reprints of published articles along with many unpublished papers), but also in the very abundant, specialized existing literature, the principal references of which are located at the end of this book.

The initial chapter introduces the principal mathematical concepts needed to characterize fractal structures. The next two chapters are given over to fractal geometries found in nature; the division of these two chapters is intended to help the presentation. Chapter 2 concerns those structures which may extend to enormous sizes (galaxies, mountainous reliefs, etc.), while Chap. 3 explains those fractal structures studied by materials physicists. This classification is obviously too rigid; for example, fractures generate similar structures ranging in size from several microns to several hundreds of meters.

In these two chapters devoted to fractal geometries produced by the physical world, we have introduced some very general models. Thus fractional Brownian motion is introduced to deal with reliefs, and percolation to deal with disordered media. This approach, which may seem slightly unorthodox seeing that these concepts have a much wider range of application than the examples to which they are attached, is intended to lighten the mathematical part of the subject by integrating it into a physical context.

Chapter 4 concerns growth models. These display too great a diversity and richness to be dispersed in the course of the treatment of the various phenomena described.

Finally, Chap. 5 introduces the dynamic aspects of transport in fractal media. Thus it completes the geometric aspects of dynamic phenomena described in the previous chapters.

I would like to thank my colleagues Pierre Collet, Eric Courtens, François Devreux, Marie Farge, Max Kolb, Roland Lenormand, Jean-Marc Luck, Laurent Malier, Jacques Peyrière, Bernard Sapoval, and Richard Schaeffer, for the many discussions which we have had during the writing of this book. I thank Benoît Mandelbrot for the many improvements he has suggested throughout this book and for agreeing to write the preface. I am especially grateful to Etienne Guyon, Jean-Pierre Hulin, Pierre Moussa, and Michel Rosso for all the remarks and suggestions that they have made to me and for the time they have spent in checking my manuscript. Finally, I would like to thank Marc Donnart and Suzanne Gouyet for their invaluable assistance during the preparation of the final version.

The success of the French original version published by Masson, has motivated Masson and Springer to publish the present English translation. I am greatly indebted to them. I acknowledge Dr. David Corfield who carried out this translation and Dr. Clarissa Javanaud and Prof. Eugene Stanley for many valuable remarks upon the final translation. During the last four years, the use of fractals has widely spread in various fields of science and technology, and some new approaches (such as wavelets transform) or concepts (such as scale relativity) have appeared. But the essential of fractal knowledge was already present at the end of the 1980s.

Palaiseau, July 1995

CHAPTER 1

Fractal Geometries

1.1 Introduction

The end of the 1970s saw the idea of *fractal geometry* spread into numerous areas of physics. Indeed, the concept of fractal geometry, introduced by B. Mandelbrot, provides a solid framework for the analysis of natural phenomena in various scientific domains. As Roger Pynn wrote in *Nature*, "If this opinion continues to spread, we won't have to wait long before the study of fractals becomes an obligatory part of the university curriculum."

The *fractal* concept brings many earlier mathematical studies within a single framework. The objects concerned were invented at the end of the 19th century by such mathematicians as Cantor, Peano, etc. The term *"fractal"* was introduced by B. Mandelbrot (fractal, i.e., that which has been infinitely divided, from the Latin *"fractus,"* derived from the verb *"frangere,"* to break). It is difficult to give a precise yet general definition of a fractal object; we shall define it, following Mandelbrot, as a set which shows irregularities on all scales.

Fundamentally it is its *geometric* character which gives it such great scope; fractal geometry forms the missing complement to Euclidean geometry and crystalline symmetry.[1] As Mandelbrot has remarked, clouds are not spheres, nor mountains cones, nor islands circles and their description requires a different geometrization.

As we shall show, the idea of fractal geometry is closely linked to properties invariant under change of scale: a fractal structure is the same *"from near or from far."* The concepts of self-similarity and scale invariance appeared independently in several fields; among these, in particular, are critical phenomena and second order phase transitions.[2] We also find fractal geometries in particle trajectories, hydrodynamic lines of flux, waves, landscapes, mountains, islands and rivers, rocks, metals, and composite materials, plants, polymers, and gels, etc.

---

[1] We must, however, add here the recent discoveries about quasicrystalline symmetries.

[2] We shall not refer here to the wide and fundamental literature on critical phenomena, renormalization, etc.

Many works on the subject have been published in the last 10 years. Basic works are less numerous: besides his articles, B. Mandelbrot has published general books about his work (Mandelbrot, 1975, 1977, and 1982); the books by Barnsley (1988) and Falconer (1990) both approach the mathematical aspects of the subject. Among the books treating fractals within the domain of the physical sciences are those by Feder (1988) and Vicsek (1989) (which particularly concentrates on growth phenomena), Takayasu (1990), or Le Méhauté (1990), as well as a certain number of more specialized (Avnir, 1989; Bunde and Havlin, 1991) or introductory monographs on fractals (Sapoval, 1990). More specialized reviews will be mentioned in the appropriate chapters.

1.2 The notion of dimension

A common method of measuring a length, a surface area or a volume consists in covering them with boxes whose length, surface area or volume is taken as the unit of measurement (Fig. 1.2.1). This is the principle which lies behind the use of multiple integration in calculating these quantities.

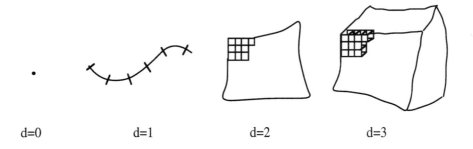

d=0   d=1   d=2   d=3

Fig. 1.2.1. Paving with lines, surfaces, or volumes.

If ε is the side (standard length) of a box and d its Euclidean dimension, the measurement obtained is

$$\mathcal{M} = N \, \varepsilon^d = N\mu,$$

where μ is the unit of measurement (length, surface area, or volume in the present case, mass in other cases). Cantor, Carathéodory, Peano, etc. showed that there exist pathological objects for which this method fails. The measurement above must then be replaced, for example, by the α-dimensional Hausdorff measure. This is what we shall now explain.

The length of the Brittany's coastline

Imagine that we would like to apply the preceding method to measure the length, between two fixed points, of a very jagged coastline such as that of

Brittany.[3] We soon notice that we are faced with a difficulty: the length L depends on the chosen unit of measurement ε and increases indefinitely as ε decreases (Fig. 1.2.2)!

Fig. 1.2.2. Measuring the length of a coastline in relation to different units.

For a standard unit ε_1 we get a length $N_1 \varepsilon_1$, but a smaller standard measure, ε_2, gives a new value which is larger,

$$L(\varepsilon_1) = N_1 \varepsilon_1$$
$$L(\varepsilon_2) = N_2 \varepsilon_2 \neq L(\varepsilon_1)$$
$$\cdots$$

and this occurs on scales going from several tens of kilometers down to a few meters. L.F. Richardson, in 1961, studied the variations in the approximate length of various coastlines and noticed that, very generally speaking, over a

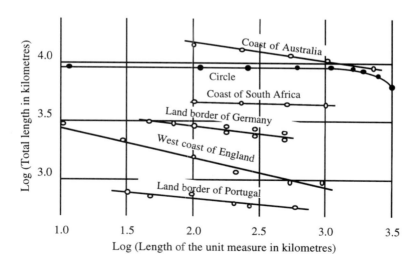

Fig. 1.2.3 Measurements of the lengths of various coastlines and land borders carried out by Richardson (1961)

---

[3] See the interesting preface of J. Perrin (1913) in *Atoms*, Constable (London).

large range of $\mathcal{L}(\varepsilon)$, the length follows a power law[4] in ε,

$$\mathcal{L}(\varepsilon) = N(\varepsilon)\,\varepsilon \propto \varepsilon^{-\rho}.$$

Figure 1.2.3 shows the behavior of various coastlines as functions of the unit of measurement. We can see that for a "normal" curve like the circle, the length remains constant ($\rho = 0$) when the unit of measurement becomes small enough in relation to the radius of curvature. The dimension of the circle is of course $D = 1$ (and corresponds to $\rho = 0$). The other curves display a positive exponent ρ so that their length grows indefinitely as the standard length decreases: it is impossible to give them a precise length, they are said to be nonrectifiable.[5] Moreover, these curves also prove to be nondifferentiable.

The exponent $(1+\rho)$ of $1/N(\varepsilon)$ defined above is in fact the "*fractal dimension*" as we shall see below. This method of determining the fractal size by covering the coast line with discs of radius ε is precisely the one used by Pontrjagin and Schnirelman (1932) (Mandelbrot, 1982, p. 439) to define the *covering dimension*. The idea of defining the dimension on the basis of a covering ribbon of width 2ε had already been developed by Minkowski in 1901. We shall therefore now examine these methods in greater detail.

Generally speaking, studies carried out on fractal structures rely both on those concerning nondifferentiable functions (Cantor, Poincaré, and Julia) and on those relating to the measure (dimension) of a closed set (Bouligand, Hausdorff, and Besicovitch).

1.3 Metric properties: Hausdorff dimension, topological dimension

Several definitions of fractal dimension have been proposed. These mathematical definitions are sometimes rather formal and initially not always very meaningful to the physicist. For a given fractal structure they usually give the same value for the fractal dimension, but this is not always the case. With some of these definitions, however, the calculations may prove easier or more precise than with others, or better suited to characterize a physical property.

Before giving details of the various categories of fractal structures, we shall give some mathematical definitions and various methods for calculating dimensions; for more details refer to Tricot's work (Tricot, 1988), or to Falconer's books (Falconer, 1985, 1990).

First, we remark that to define the dimension of a structure, this structure must have a notion of distance (denoted |x-y|) defined on it between any two of its points. This hardly poses a problem for the structures provided by nature.

---

[4] The commonly used notation '$\propto$' means 'varies as': a $\propto$ b means precisely that the ratio a/b asymptotically tends towards a nonzero constant.

[5] A part of a curve is rectifiable if its length can be determined.

We should also mention that in these definitions there is always a passage to the limit ε⁻0. For the actual calculation of a fractal dimension we are led to discretize (i.e., to use finite basic lengths ε): the accuracy of the calculation then depends on the relative lengths of the unit ε, and that of the system (Sec. 1.4.4).

1.3.1 The topological dimension d_T

If we are dealing with a geometric object composed of a set of points, we say that its fractal dimension is $d_T = 0$; if it is composed of line elements, $d_T = 1$, surface elements $d_T = 2$, etc.

> "Composed" means here that the object is locally homeomorphic to a point, a line, a surface. The topological dimension is invariant under invertible, continuous, but not necessarily differentiable, transformations (homeomorphisms). The dimensions which we shall be speaking of are invariant under differentiable transformations (dilations).

A fractal structure possesses a fractal dimension strictly greater than its topological dimension.

1.3.2 The Hausdorff–Besicovitch dimension, or covering dimension: dim(E)

The first approach to finding the dimension of an object, E, follows the usual method of covering the object with boxes (belonging to the space in which the object is embedded) whose measurement unit $\mu = \varepsilon^{d(E)}$, where $d(E)$ is the Euclidean dimension of the object. When $d(E)$ is initially unknown, one possible solution takes $\mu = \varepsilon^\alpha$ as the unit of measurement for an unknown exponent α. Let us consider, for example, a square ($d = 2$) of side L, and cover it with boxes of side ε. The measure is given by $\mathcal{M} = N\mu$, where N is the number of boxes, hence $N = (L/\varepsilon)^d$. Thus,

$$\mathcal{M} = N \varepsilon^\alpha = (L/\varepsilon)^d \varepsilon^\alpha = L^2 \varepsilon^{\alpha-2}$$

If we try $\alpha = 1$, we find that $\mathcal{M} \to \infty$ when $\varepsilon \to 0$: the "length" of a square is infinite. If we try $\alpha = 3$, we find that $\mathcal{M} \to 0$ when $\varepsilon \to 0$: the "volume" of a square is zero. The surface area of a square is obtained only when $\alpha = 2$, and its dimension is the same as that of a surface $d = \alpha = 2$.

The fact that this method can be applied for any real α is very interesting as it makes possible its generalization to noninteger dimensions.

We can formalize this measure a little more. First, as the object has no specific shape, it is not possible, in general, to cover it with identical boxes of side ε. But the object E may be covered with balls V_i whose diameter (diam V_i) is less than or equal to ε. This offers more flexibility, but requires that the inferior limit of the sum of the elementary measures be taken as $\mu = (\text{diam } V_i)^\alpha$.

Therefore, we consider what is called the α–covering measure (Hausdorff, 1919; Besicovitch, 1935) defined as follows:

$$m^\alpha(E) = \lim_{\varepsilon \to 0} \inf\{ \Sigma(\text{diam } V_i)^\alpha : \cup V_i \supset E, \text{ diam } V_i \leq \varepsilon\}, \qquad (1.3-1)$$

and we define the *Hausdorff* (or Hausdorff–Besicovitch) *dimension*: dim E by

$$\begin{aligned} \dim E &= \inf \{ \alpha : m^\alpha(E) = 0 \} \\ &= \sup \{ \alpha : m^\alpha(E) = \infty\}. \end{aligned} \qquad (1.3-2)$$

The Hausdorff dimension is the value of α for which the measure jumps from zero to infinity. For the value $\alpha = \dim E$, this measure may be anywhere between zero and infinity.

The function $m^\alpha(E)$ is monotone in the sense that if a set F is included in E, $E \supset F$, then $m^\alpha(E) \geq m^\alpha(F)$ whatever the value of α.

1.3.3 The Bouligand–Minkowski dimension

We can also define a dimension known as the Bouligand–Minkowski dimension (Bouligand, 1929; Minkowski, 1901), denoted $\Delta(E)$. Here are some methods of calculating $\Delta(E)$:

The Minkowski sausage (Fig. 1.3.1)

Let E be a fractal set embedded in a d-dimensional Euclidean space (more precisely E is a closed subset of R^d). Now let $E(\varepsilon)$ be the set of points in R^d at a distance less than ε from E. $E(\varepsilon)$ now defines a Minkowski sausage: it is also called a thickening or dilation of E as in image analysis. It may be defined as the union

$$E(\varepsilon) = \underset{x \in E}{\cup} B_\varepsilon(x),$$

where $B_\varepsilon(x)$ is a ball of the d-dimensional Euclidean space, centered at x and of radius ε. We calculate,

$$\Delta(E) = \lim_{\varepsilon \to 0} \left(d - \frac{\log \text{Vol}_d[E(\varepsilon)]}{\log \varepsilon} \right), \qquad (1.3-3)$$

where Vol_d simply represents the volume in d dimensions (e.g., the usual length, surface area, or volume). If the limit exists, $\Delta(E)$ is, by definition, the *Bouligand–Minkowski dimension*.

Naturally, we recover from this the usual notion of dimension: let us take as an example a line segment of length L. The associated Minkowski sausage has as volume $\text{Vol}_d(E)$,

$$\begin{aligned} &\text{in } d = 2: &&2\varepsilon L + \pi\varepsilon^2, \\ &\text{in } d = 3: &&\pi\varepsilon^2 L + (4\pi/3)\varepsilon^3, \end{aligned}$$

Fig. 1.3.1. Minkowski sausage or thickening of a curve E.

so that neglecting higher orders in ε, $\mathrm{Vol}_d(E) \propto \varepsilon^{d-1}$.

In general terms we have:

If E is a point:	$\mathrm{Vol}_d(E) \propto \varepsilon^d$,	$\Delta(E) = 0$.
If E is a rectifiable arc:	$\mathrm{Vol}_d(E) \propto \varepsilon^{d-1}$,	$\Delta(E) = 1$.
If E is a k-dimensional ball:	$\mathrm{Vol}_d(E) \propto \varepsilon^{d-k}$,	$\Delta(E) = k$.

In practice, $\Delta(E)$ is obtained as the slope of the line of least squares of the set of points given by the plane coordinates,
$$\{ \ \log 1/\varepsilon, \quad \log \mathrm{Vol}_d [E(\varepsilon)/\varepsilon^d] \ \}.$$
This method is easy to use. The edge effects (like those obtained above in measuring a segment of length L) lead to a certain inaccuracy in practice (i.e., to a curve for values of ε which are not very small).

The box-counting method (Fig. 1.3.2)

This is a very useful method for many fractal structures. Let $N(\varepsilon)$ be the number of boxes of side ε covering E:

$$\Delta(E) = \lim_{\varepsilon \to 0} \left(\frac{\log N(\varepsilon)}{-\log \varepsilon} \right) \tag{1.3-4}$$

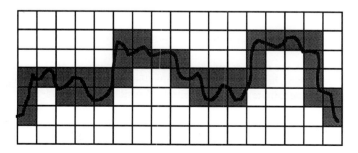

Fig.1.3.2. Measurement of the dimension of a curve by the box-counting method.

The box-counting method is commonly used, particularly for self-affine structures (see Sec. 2.2.5).

    The dimension of a union of sets is equal to the largest of the dimensions of these sets: $\Delta(E \cup F) = \max \{\Delta(E), \Delta(F)\}$.
    The limit $\Delta(E)$ may depend on the choice of paving. If there are two different limits Sup and Inf, the Sup limit should be taken.

The disjointed balls method (Fig 1.3.3)

Let $N(\varepsilon)$ be the *maximum* number of disjoint balls of radius ε centered on the set E: then

$$\Delta(E) = \lim_{\varepsilon \to 0} \log N(\varepsilon) \,/\, |\log \varepsilon| \,. \qquad (1.3\text{-}5)$$

This method is rarely used in practice.

Fig.1.3.3. Measuring the dimension of a curve by the disjointed balls method.

The dividers' method (Richardson, 1960)

This is the method we described earlier (Fig. 1.2.2).

Let $N(\varepsilon)$ be the number of steps of length ε needed to travel along E:

$$\Delta(E) = \lim_{\varepsilon \to 0} \log N(\varepsilon) \,/\, |\log \varepsilon| \qquad (1.3\text{-}6)$$

    Notice that all the methods give the same fractal dimension, $\Delta(E)$, when it exists (see Falconer, 1990), because we are in a finite dimensional Euclidean space. This is no longer true in an infinite dimensional space, (function space, etc.).

1.3.4 The packing dimension [or Tricot dimension: Dim (E)]

    Unlike the Hausdorff–Besicovitch dimension, which is found using the α-dimensional Hausdorff measure, the box-counting dimension $\Delta(E)$ is not defined in terms of measure. This may lead to difficulties in certain theoretical developments. This problem may be overcome by defining the packing dimension, following similar ideas to those of the α-dimensional Hausdorff measure (Falconer, 1990). Let $\{V_i\}$ be a collection of disjoint balls, and

$$\mathcal{P}_0^\alpha(E) = \lim_{\varepsilon \to 0} \sup\{\, \Sigma(\text{diam } V_i)^\alpha, \text{ diam } V_i \le \varepsilon\}.$$

As this expression is not always a measure we must consider

$$\mathcal{P}^\alpha(E) = \inf\{\, \mathcal{P}_0^\alpha(E_i) : \bigcup_{i=1}^{\infty} E_i \supset E\} \ .$$

The packing dimension is defined by the following limit:

$$\text{Dim } E = \sup\{\alpha : \mathcal{P}^\alpha(E) = \infty\} = \inf\{\alpha : \mathcal{P}^\alpha(E) = 0\}, \qquad (1.3\text{-}7a)$$

alternatively, according to the previous definitions:

$$\text{Dim } E = \inf\{\sup \Delta(E_i) : \cup E_i \supset E\}. \qquad (1.3\text{-}7b)$$

The following inequalities between the various dimensions defined above are always true:

$$\dim E \le \text{Dim } E \ \le \Delta(E)$$
$$\dim E + \dim F \ \le \dim E \otimes F$$
$$\le \dim E + \text{Dim } F$$
$$\le \text{Dim } E \otimes F \le \text{Dim } E + \text{Dim } F.$$

Notice that for multifractals box-counting dimensions are in practice rather Tricot dimensions.

Other methods of calculation have been proposed by Tricot (Tricot, 1982) which could prove attractive in certain situations. Without entering into the details, we should also mention the *method of structural elements*, the *method of variations* and the *method of intersections*.

Theorem: If there exists a real D and a finite positive measure μ such that for all $x \in E$, ($B_r(x)$ being the ball of radius r centered at x),

$$\log \mu[B_r(x)]/\log r \to D, \text{ then}$$

$$D = \dim E. \qquad (1.3\text{-}8)$$

D is also called the *mass dimension*. If the convergence is uniform on E, then

$$D = \dim(E) = \Delta(E). \qquad (1.3\text{-}9)$$

This theorem does not always apply: dim E = 0 for a denumerable set, while for the Bouligand–Minkowski dimension $\Delta(E) \ne 0$.

In practice, Mandelbrot has popularized the Hausdorff–Besicovitch dimension or mass dimension (as the measure is very often a mass), dim E, which turns out to be one of the simpler and more understandable dimensions (although not always the most appropriate) for the majority of problems in physics *when the above theorem applies*.

So we now have the following relation giving the mass inside a ball of radius r,

$$\mathcal{M} = \mu(B_r(x)) \propto r^D,$$

(1.3-10)

where the center x of the ball B is inside the fractal structure E.

We shall of course take the physicist's point of view and not burden ourselves, at first, with too much mathematical rigor. The *fractal dimension will in general be denoted* D and, in the cases considered, we shall suppose that, unless specified otherwise, the existence theorem applies and therefore that the fractal dimension is the same for all the methods described above.

Units of measure

The above relation can often be written in the form of a dimensionless equation, by introducing the unit of length ε_u and volume $(\varepsilon_u)^d$ or mass $\rho_u = (\varepsilon_u)^d \rho$ (by assuming a uniform density ρ over the support):

$$\frac{\mathcal{V}}{(\varepsilon_u)^d} \quad \text{or} \quad \frac{\mathcal{M}}{\rho_u} \propto \left(\frac{r}{\varepsilon_u}\right)^D.$$

Examples of this for the Koch curve and the Sierpinski gasket will be given later in Sec. 1.4.1. In this case the unit of volume is that of a space with dimension equal to the topological dimension of the geometric objects making up the set (see Sec. 1.3.1).

> From a strictly mathematical point of view the term "dimension" should be reserved for sets. For measures, we can think of the set covered by a uniform measure. However, we can define the dimension of a measure by
>
> $$\dim(\mu) = \inf \{ \dim(A), \mu(A^c)=0 \},$$
>
> A being a measurable set and A^c its complement. This dimension is often strictly less than the dimension of the support. This happens with the information dimension described in Sec. 1.6.2.
>
> For objects with a different scaling factor in different spatial directions, the box-counting dimension differs from the Hausdorff dimension (see Fig. 2.2.8).

Having defined the necessary tools for studying fractal structures, it is now time to get to the heart of the matter by giving the first concrete examples of fractals.

1.4 Examples of fractals

1.4.1 Deterministic fractals

Some fractal structures are constructed simply by using an iterative process consisting of an initiator (initial state) and a generator (iterative operation).

The triadic Von Koch curve (1904)

Each segment of length ε is replaced by a broken line (*generator*), composed of four segments of length $\varepsilon/3$, according to the following recurrence relation:

(generator)

At iteration zero, we have an *initiator* which is a segment in the case of the triadic Koch curve, or an equilateral triangle in the case of the Koch island. If the initiator is a segment of horizontal length L, at the first iteration (the curve coincides with the generator) the base segments will have length $\varepsilon_1 = L/3$;

at the second iteration they will have length $\varepsilon_2 = L/9$ as each segment is again replaced by the generator, then $\varepsilon_3 = L/3^3$ at the third iteration

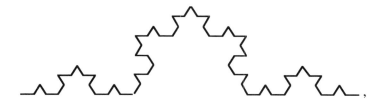

and so on. The relations giving the length $\mathcal{L}$ of the curve are thus

$$\varepsilon_1 = L/3 \;\rightarrow\; \mathcal{L}_1 = 4\,\varepsilon_1$$
$$\varepsilon_2 = L/9 \;\rightarrow\; \mathcal{L}_2 = 16\,\varepsilon_2$$
$$\dots$$
$$\varepsilon_n = L/3^n \;\rightarrow\; \mathcal{L}_n = 4^n\,\varepsilon_n$$

by eliminating n from the two equations in the last line, the length $\mathcal{L}_n$ may be written as a function of the measurement unit ε_n

$$\mathcal{L}_n = L^D\,(\varepsilon_n)^{1-D} \quad \text{where } D = \log 4 \,/\, \log 3 = 1.2618\dots$$

For a *fixed unit length* ε_n, $\mathcal{L}_n$ grows as the Dth power of the size L of the curve. Notice that here again we meet the exponent $\rho = D-1$ of ε_n, which we first met in Sec. 1.2 (Richardson's law) and which shows the divergence of $\mathcal{L}_n$ as $\varepsilon_n \to 0$.

At a given iteration, the curve obtained is not strictly a fractal but according to Mandelbrot's term a "prefractal". A fractal is a mathematical object obtained in the limit of a series of prefractals as the number of iterations n tends to infinity. In everyday language, prefractals are often both loosely called "fractals".

The previous expression is the first example given of a scaling law which may be written

$$\mathcal{L}_n / \varepsilon_n = f(L / \varepsilon_n) = (L / \varepsilon_n)^D \quad . \tag{1.4-1}$$

A scaling law is a relation between different *dimensionless* quantities describing the system, (the relation here is a simple power law). Such a law is generally possible only when there is a single independent unit of length in the object (here ε_n).

A structure associated with the Koch curve is obtained by choosing an equilateral triangle as initiator. The structure generated in this way is the well-known *Koch island* (see Fig. 1.4.1).

Fig. 1.4.1. Koch island after only three iterations. Its coastline is fractal, but the island itself has dimension 2 (it is said to be a surface fractal).

Simply by varying the generator, the Koch curve may be generalized to give curves with fractal dimension $1 \le D \le 2$. A straightforward example is provided by the modified Koch curve whose generator is

and whose fractal dimension is $D = \log 4 / \log [2 + 2 \sin(\alpha/2)]$. Notice that in the limit $\alpha = 0$ we have $D = 2$, that is to say a curve which fills a triangle. It is not exactly a curve as it has an infinite number of multiple points. But the construction can be slightly modified to eliminate them. The dimension $D = 2$

(= log 9/log 3) is also obtained for the *Peano curve* (Fig. 1.4.2) (which is dense in a square) whose generator is formed from 9 segments with a change by a factor 3 in the linear dimension, i.e.,

This gives after the first three iterations,

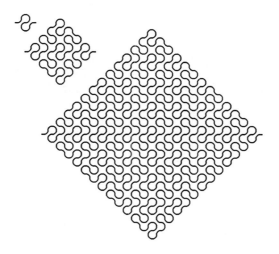

Fig. 1.4.2. First three iterations of the Peano curve (for graphical reasons the scale is simultaneously dilated at each iteration by a factor of 3). The Peano curve is dense in the plane and its fractal dimension is 2.

This construction has also been modified (rounding the angles) to eliminate double points.

The von Koch and Peano curves are as their name indicates: curves, that is, their *topological dimension* is

$$d_T = 1.$$

Practical determination of the fractal dimension using the mass-radius relation

As mentioned earlier, a method which we shall be using frequently to determine fractal dimensions[6] consists in calculating the mass of the structure within a ball of dimension d centered on the fractal. If the embedding space is d-dimensional, and of radius R, then

$$\mathcal{M} \propto R^D.$$

The measure here is generally a mass, but it could equally well be a "surface area" or any other scalar quantity attached to the support (Fig. 1.4.3).

---

[6] The box-counting method will also be frequently used.

In the case of the Koch curve, we could check to see that $D = \log 4 / \log 3$, as is the case for the different methods shown above. Notice that if the ε_n are not chosen in the sequence $\varepsilon_n = L/3^n$, the calculations prove much more complicated, but the limit as $\varepsilon \to 0$ still exists and gives D.

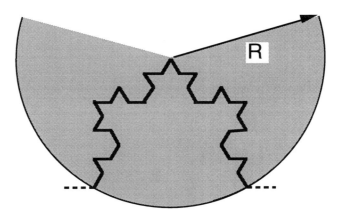

Fig. 1.4.3. Measuring the fractal dimension of a Koch curve using the relationship bet-ween mass and radius. If each segment represents a unit of "fractal surface area" (1 cm^D, say), the "surface area" above is equal to 2 cm^D when R = 1 cm, 8 cm^D when R = 3 cm.

In very general terms $\mathcal{M}$ has the form,

$$\mathcal{M} = A(R)\ R^D$$

where $A(R) = A_0 + A_1\ R^{-\Omega} + \dots$ tends to a constant A_0 as $R \to \infty$. When the coefficients $A_1, \dots$ are nonzero (which is not the case for the examples in this chapter), $A(R)$ is called the scaling law correction.

Direct determination of the fractal dimension and the multiscale case

The fractal dimension D may be found directly from a single iteration *if the limit structure is known to be a fractal*. If a fractal structure of size L with mass $\mathcal{M}(L) = A(L)\ L^D$ gives after iteration k elements of size L/h, we then have an implicit relation in D:

$$\mathcal{M}(L) = k\ \mathcal{M}(L/h), \quad \text{hence}\quad A(L)\ L^D = k\ A(L/h)\ (L/h)^D.$$

D is thus determined asymptotically ($L \to \infty$) by noticing that

$$A(L/h)\ /\ A(L) \to 1 \text{ as } L \to \infty. \text{ Hence } k\ (1/h)^D = 1.$$

For example, the Koch curve corresponds to $k = 4$ and $h = 3$. Moreover, $A(L)$ is independent of L here.

Later on we shall meet *multiscale fractals*, giving at each iteration k_i elements of size L/h_i $(i = 1, \dots, n)$. Thus

$$\mathcal{M}(L) = k_1 \, \mathcal{M}(L/h_1) + k_2 \, \mathcal{M}(L/h_2) + ... + k_n \, \mathcal{M}(L/h_n),$$

which means that the mass of the object of linear size L is the sum of k_i masses of similar objects of size L/h_i. Thus,

$$k_1 (1/h_1)^D + k_2 (1/h_2)^D ... + k_n (1/h_n)^D = 1, \qquad (1.4\text{-}2)$$

which determines D.

Cantor sets

These are another example of objects which had been much studied before the idea of fractals was introduced. The following Cantor set is obtained by iteratively deleting the central third of each segment:

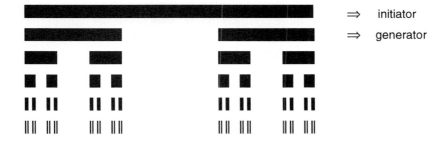

$\Rightarrow$ initiator

$\Rightarrow$ generator

Fig. 1.4.4. Construction of the first five iterations of a Cantor set. In order to have a clearer representation and to introduce the link between measure and set, the segments have been chosen as bars of fixed width (Cantor bars), consequently representing a uniform density distributed over the support set (uniform measure, see also Sec.1.6.3). In this way the fractal dimension and the mass dimension are identified.

Five iterations are shown in Fig. 1.4.4.

The fractal dimension of this set is

$$D = \log 2/ \log 3 = 0.6309...$$

For Cantor sets we have $0 < D < 1$: it is said to be a "dust." As it is composed only of points, its *topological dimension* is $d_T = 0$.

To demonstrate the fact that the fractal dimension by itself does not uniquely characterize the object, we now construct a second Cantor set with the same fractal dimension but a different spatial structure (Fig. 1.4.5): at each iteration, each element is divided into four segments of length 1/9, which is equivalent to uniformly spacing the elements of the second iteration of the previous set. In fact these two sets differ by their lacunarity (cf. Sec. 1.5.3), that is, by the distribution of their empty regions.

Fig. 1.4.5. Construction of the first two iterations of a different Cantor set having the same fractal dimension.

Mandelbrot–Given curve

Iterative deterministic processes have shown themselves to be of great value in the study of the more complex fractal structures met with in nature, since their iterative character often enables an exact calculation to be made. The Mandelbrot-Given curve (Mandelbrot and Given, 1984) is an instructive example of this as it simulates the current conducting cluster of a network of resistors close to their conductivity threshold (a network of resistors so many of which are cut that the network barely conducts). It is equally useful for understanding multifractal structures (see Fig. 1.4.6). We shall take this up again in Sec 5.2.2 (hierarchical models) as it is a reasonable model for the "backbone" of the infinite percolation cluster (Fig. 3.1.8).

The generator and first two iterations are as follows:

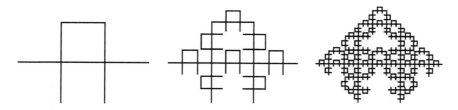

Fig. 1.4.6. Construction of the first three iterations of a Mandelbrot–Given set. This fractal has a structure reminiscent of the percolation cluster which plays an important role in the description of disordered media (Sec. 3.1).

The vertical segments of the generator are slightly shortened to avoid double points. The fractal dimension (neglecting the contraction of the vertical segments) is $D = \log 8 / \log 3 \cong 1.89...$.

"Gaskets" and "Carpets"

These structures are frequently used to carry out exact, analytic calculations of various physical properties (conductance, vibrations, etc.).

Sierpinski gasket

Fig. 1.4.7. Iteration of the Sierpinski gasket composed of full triangles (the object is made up only of those parts left coloured black).

The scaling factor of the iteration is 2, while the mass ratio is 3 (see Fig. 1.4.7). The corresponding fractal dimension is thus

$$D = \log 3/\log 2 = 1.585...$$

The Sierpinski gasket generated by the edges only is also often used (see Fig. 1.4.8). It clearly has the same fractal dimension $D = \log 3/\log 2 = 1.585...$

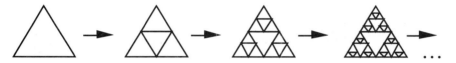

Fig. 1.4.8. Iteration of the Sierpinski gasket composed of sides of triangles.

The two structures can be shown to "converge" asymptotically towards one other, in the sense of the Hausdorff distance (see, e.g., Barnsley, 1988).

Sierpinski carpet

Fig. 1.4.9. Iteration of a Sierpinski carpet.

The scaling factor is 3 and the mass ratio (black squares) is 8 (see Fig. 1.4.9). Hence,

$$D = \log 8/\log 3 = 1.8928...$$

Other examples

Examples of deterministic fractal structures constructed on the basis of the Sierpinski gasket and carpet can be produced endlessly. These geometries can prove very important for modelling certain transport problems in porous objects or fractal electrodes. Here are a couple of three-dimensional examples, the 3d gasket and the so-called Menger sponge (Fig. 1.4.10).

3d gasket (Mandelbrot)	*Menger sponge*
$D = \log 4/\log 2 = 2$	$D = \log 20/\log 3 \cong 2.73$

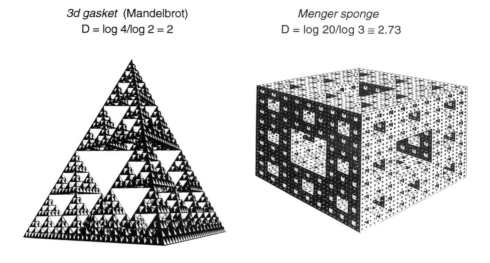

Fig. 1.4.10. Three-dimensional Sierpinski gasket composed of full tetrahedra (above left); Menger sponge (above right). (Taken from Mandelbrot, 1982.)

This again illustrates the fact that fractal dimension alone does not characterize the object: the fractal dimension of the three-dimensional gasket is equal to two, as is that of the Peano curve.

Nonuniform fractals

Another possible type of fractal structure relies on the simultaneous use of several dilation scales. Here is an example of such a structure, obtained by deterministic iteration using factors 1/4 and 1/2 (Fig. 1.4.11). This structure is clearly fractal and its dimension D is determined by one iteration, as before, (from Eq. 1.4.2):

$$4\,(L/4)^D + (L/2)^D = L^D$$

$$\text{hence} \quad D = \frac{\log(1+\sqrt{17}\,)}{\log 2} - 1 \quad.$$

In reality, its support distribution is more complex than the fractals described above: in fact it is multifractal. Multifractal measures will be studied in greater detail in Sec. 1.6.

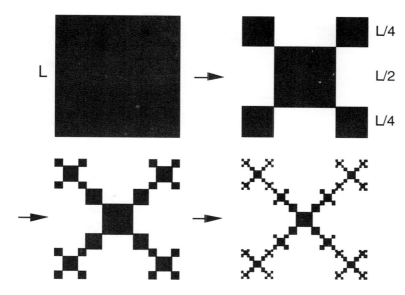

Fig. 1.4.11. Construction of a nonuniform deterministic fractal: here with two scales of contraction.

1.4.2 Random fractals

Up to now only examples of deterministic (also called "exact") fractals have been given, but random structures can easily be built. In these structures the recurrence defining the hierarchy is governed by one or more probabilistic laws which fix the choice of which generator to apply at each iteration.

Homogeneous fractals

A random fractal is homogeneous when the structure's volume (or mass) is distributed uniformly at each hierarchical level, that is, the different generators used to construct the fractal keep the same mass ratio from one level to the next.

So, from the recurrence:

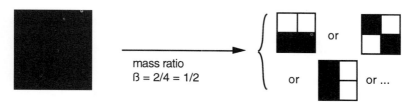

Fig. 1.4.12-a. Random fractal generator.

the following random fractal may be constructed:

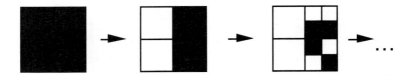

Fig. 1.4.12-b. Random fractal generated by the previous generator.

The corresponding fractal dimension is

$$D = d + \log \beta / \log 2 = 1.$$

Finding D from a single iteration, we have $2^d\beta$ new elements, each of size 1/2 at each iteration, thus $2^d\beta (1/2)^D = 1$.

Heterogeneous fractals

The mass ratio ß may itself vary: a fractal constructed in this way is said to be heterogeneous (Figs. 1.4.13a and 1.4.13b). This type of fractal can be used as a basis for modeling turbulence (see Sec. 2.3).

Starting with a recurrence relation, with a given distribution of ß,

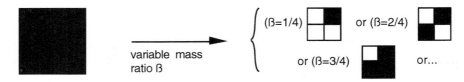

Fig. 1.4.13-a. Generator of a heterogeneous random fractal.

it is possible to build up heterogeneous fractal structures:

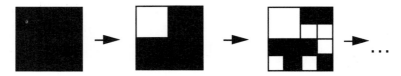

Fig. 1.4.13-b. Heterogeneous random fractal generated by the previous fractal.

whose dimension is given by $\langle \mathcal{M}(L) \rangle \propto L^D$. Hence,

$$D = d + \log \langle ß \rangle / \log 2$$

Random fractals are, with some notable exceptions, almost the only ones found in nature; their fractal properties (scale invariance, see Sec. 1.4.3) bear on the statistical averages associated with the fractal structure.

Example: Fig 1.4.14 below shows a distribution of disks ,the positions of whose centers follow a Poisson distribution, and whose radii are randomly distributed according to a probability density, $P(R>r) = Q\,r^{-\alpha}$; the larger the value of α, the higher is the frequency of smaller disks, and the further the fractal dimension of the black background is from 2. Such a distribution of disks could belong to lunar craters seen from above (projection) or holes in a piece of Emmenthal cheese! We shall be returning to this model later.

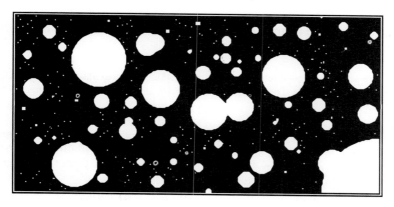

Fig. 1.4.14 Random fractal of discs whose sizes are distributed according to a power law.

In a similar way any number of fractal structures may be thought up. Examples are given in Mandelbrot (1982).

Let us now return to the basic properties of the structures we have just constructed.

1.4.3 Scale invariance

Scale invariance is also called *invariance under internal similarity* or sometimes *dilation invariance*. It is a feature which allows one to detect a fractal structure simply by looking at it: the object appears similar to itself "from near as from far," that is, whatever the scale. Naturally the eye is inadequate and a more refined analysis is required. In previous examples this invariance came from the fact that an iterative structure is such that its mass obeys a homothetic relation of the form[7] (L large)

$$\mathcal{M}(bL) = \lambda \mathcal{M}(L),$$

so that by dilating the linear dimensions of a volume by a factor b, the mass of the matter contained in this volume will be multiplied by a factor λ. For

---

[7] The unit of measure must be conserved in the dilation; to avoid difficulties due to a characteristic minimum size in the physical system considered (an aggregate of particles for example), we take this size to be the unit.

ordinary surfaces or volumes $\lambda = b^d$, where d is the dimension of the object. This relation generalizes to all self-similar fractals. For the Koch curve, for example,

$$\mathcal{M}(3L) = 4\mathcal{M}(L) = 3^D \mathcal{M}(L),$$

and in the general case,

$$\boxed{\mathcal{M}(bL) = b^D \mathcal{M}(L)} \qquad (1.4\text{-}3)$$

This is a very direct method of calculating D, which is thus also *the similarity dimension*, (see also the remark p. 14).

> The scale invariance relation $\mathcal{M}(bL) = b^D \mathcal{M}(L)$ is equivalent to the mass/radius relation $\mathcal{M}(L) = A_0 L^D$. To see this, choose b = 1/L in the scaling law, giving $\mathcal{M}(L) = \mathcal{M}(1) L^D$.

We shall see that internal similarity is present in many exact or random fractals (see below), which are not generated by iteration.

Generally speaking:

> *Translational invariance* $\rightarrow$ periodic networks
> *Dilation invariance* $\rightarrow$ self-similar fractals

In practice scale invariance only works for a limited range of distances r:

$$a \ll r \ll \Lambda.$$

Λ is the macroscopic limit due to the size of the sample, correlation length, effects of gradients, etc. and a is the microscopic limit due to the lattice distance, molecular sizes, etc. When we come to discuss macroscopic structures in chapter 2 (and microscopic structures in chapter 3), we mean to say that the scales of a and Λ are macroscopic (or microscopic, respectively). Moreover, if there are corrections to the scaling law [A(r) not constant], scale invariance will only be found asymptotically (for very large r).

1.4.4 Ambiguities in practical measurements

In practice, in other words, for physical objects, we find problems in applying the methods we have just described. This is partly because, as we mentioned in the last paragraph, there is both a minimum characteristic size below which the fractal description ceases to be valid (for example, aggregates composed of small particles) and also an upper size limit for the object under consideration. But it is also because a physical phenomenon, dependent on some (dominant) parameter, only generates fractal structures on all scales for a critical value of this parameter. This value is often difficult to attain, so scaling law corrections are generally required (see remark p. 14).

The fractal dimension is obtained from the slope of the linear regression of the points with coordinates, {log 1/r, log N(r)}, with r going from the minimum characteristic size to the size of the object.

The method employed to determine the dimension (discussed by Tricot, 1982), the limited extent of a fractal dynamic (object only fractal over a scale spanning less than two orders of magnitude), and the presence of scaling law corrections, all contribute to making this method based on log-log regression sometimes imprecise. Such a representation tends to straighten any curves. The presence of slight curvature is a sign that the asymptotic régime has not been reached. Furthermore, as a point of inflexion may be taken for linear behavior, it is not always obvious, without reliable theoretical support, what is happening in this case. A good theoretical model or a dynamic that extends over a sufficiently wide range of scales is therefore indispensable. A discussion about real and apparent power laws may be found in T. A. Witten, Les Houches, 1985 (Boccara and Daoud, 1985).

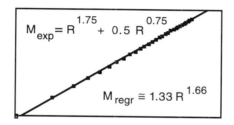

Fig. 1.4.15. Very simple example of imprecision in estimating the exponent (M_{regr}) due to corrections in the scaling law (M_{exp}). The value is found to be 1.66 instead of 1.75, in spite of the seemingly linear nature of the graph.

1.5 Connectivity properties

A fractal is insufficiently characterized by its fractal dimension D, alone. Exponents appear throughout this book with the intension of specifying the behavior of various physical quantities.

Geometrically it is interesting to consider quantities such as the *spreading*, the *ramification*, or the *lacunarity dimensions*, which characterise the properties of connectivity and of the distribution of matter within a fractal.

1.5.1 Spreading dimension, dimension of connectivity

Let us consider a fractal composed of squares lying next to each other (Fig 1.5.1). The white squares are permitted sites, the black squares are forbidden. By counting the number S(ℓ) of permitted sites accessible from a given permitted site in ℓ or less steps, where a step consists of passing to a bordering permitted site, generally, a mass–(distance in steps) relation of the form,

$$S(\ell) \propto \ell^{d_e} \tag{1.5-1}$$

is obtained. ℓ is called the *chemical distance* or *distance of connectivity* and d_e the *spreading dimension*. This dimension depends solely on the connections between the elements of the fractal structure (and not on the metric of the space in which it is embedded): it is an *intrinsic connectivity property* of the fractal. The dimension d_s is linked to the *tortuosity*: the higher the tortuosity, the more the convolutions of the object make us use detours to get from one point to another situated at a fixed distance "as the crow flies." We have the inequality $d_e \leq D$. The equality, $d_e = D$, is attained when the fractal metric corresponds to its natural metric, that is, when the Euclidean dimensions (as the crow flies) are equal or proportional to the distances obtained by staying inside the fractal (on average). This is the case for Sierpinski gaskets.

Because of the statistical invariance of the structure under any dilation, we would expect the mean quadratic Euclidean distance $R(\ell)^2$ between two sites separated by a "chemical distance" ℓ to be such that

$$\boxed{R(\ell)^2 \propto \ell^{2/d_{min}}} \tag{1.5-2}$$

d_{min} is called the chemical dimension or the dimension of connectivity. It is the fractal dimension of the shortest path (the measure being the distance in steps): we then have a (distance in steps)–radius relation given by $\ell \propto R^{d_{min}}$. Since $S(\ell) \propto R(\ell)^D$ we have

$$\boxed{d_e = \frac{D}{d_{min}}} \tag{1.5-3}$$

This only agrees with our previous statement that $d_e = D$, if $d_{min} = 1$, that is, $R \propto \ell$.

The figures represent the chemical distance $\ell = 1, 2, 3,...$ from an origin 0. The forbidden regions are in black. The shaded areas correspond to different clusters inaccessible from 0. R is the square root of the mean square distance between two points, ℓ apart .

Fig. 1.5.1. *Representation on a square network of accessible sites, chemical distance* ℓ *and visited sites* $S(\ell)$. *For example,* $S(3) = 11 = 3$ (1)+3 (2)+5 (3).

1.5.2 The ramification $\mathcal{R}$

$\mathcal{R}$ is the *smallest* number of links which must be cut to *disconnect a macroscopic part* of the object. Links should be understood as paths leading from one point of the structure to another.

Thus, for the Sierpinski gasket $\mathcal{R}$ is finite, while for the carpet $\mathcal{R}$ is infinite. Ramification plays an important role in the conduction and mechanical properties of fractals. The condition that $\mathcal{R}$ be finite is, moreover, a necessary condition for *exact* relations of the renormalization group in real space. This will be proven for the example of a Sierpinski gasket, where its vibration modes are calculated in Sec. 5.1.1 and its conductance in Sec. 5.2.1.

1.5.3 The lacunarity $\mathcal{L}$

This indicates, in some sense, how far an object is from being translationally invariant, by measuring the presence of sizeable holes in a fractal structure E.

We have seen that it is always possible to write $\mathcal{M}(R)=A(R)\,R^D$, the sole condition on A being that log A/log R→0. The distribution of holes or lacunae is consequently related to the fluctuations around the law in R^D. The lacunarity $\mathcal{L}$ is therefore defined by

$$\mathcal{L} = \text{variance (A)},$$

that is to say, the lacunarity may be calculated from the mean over E (Fig. 1.5.2),

$$\mathcal{L}(R) = (\,\langle\,\mathcal{M}(R)^2\,\rangle - \langle\,\mathcal{M}(R)\,\rangle^2\,)^{1/2} / \langle\,\mathcal{M}(R)\,\rangle\,. \qquad (1.5\text{-}4)$$

For an object such as the Cantor set, defined on p. 15, the lacunarity is periodic (in log R), as A(R) = A(bR), b = 1/3 and 1/4 for the first and second

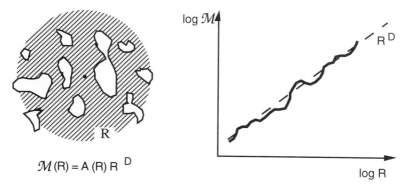

$$\mathcal{M}(R) = A(R)\,R^D$$

Fig. 1.5.2. *Effect of lacunarity on the mass relation as a function of the radius.* $\mathcal{M}(R)$ *fluctuates around the power law in* R^D.

examples, respectively. This is also true for deterministic fractals obtained by iterating a generator. The lacunarity becomes aperiodic for random fractals (Gefen et al., 1983).

1.6 Multifractal measures

Knowledge of the fractal dimension of a set (as we have seen), is insufficient to characterize its geometry, and, all the more so, any physical phenomenon occurring on this set. Thus, in a random network of fusible links, the links which melt are those through the current exceeds a certain threshold. Their distribution is supported by a set whose fractal dimension generally differs from that of the whole. Likewise, in a growth phenomenon, such as diffusion limited aggregation, (the DLA model is described in Sec. 4.2), the growth sites do not all have the same weight; some of them grow much more quickly than others. Therefore, to understand many physical phenomena, involving fractal supports, the (singular) distribution of measures associated with each point of the support must be characterized. These measures, scalar quantities, may correspond to concentrations, currents, electrical or chemical potentials, probabilities of reaching each point of the support, pressures, dissipations, etc.

Intuitive approach to multifractality

Let us take as an example the distribution of diamonds over the surface of the earth. We shall suppose that the statistics governing this distribution has certain similarities with the one in Fig. 1.4.14 (a little too optimistic !): that is, we suppose that the distribution of diamonds is fractal, that it is not homogeneous, and that there exist very few regions where large stones occur: the majority of places on the earth's surface containing only traces of diamonds. The information we can draw from knowing the fractal dimension is global: if it is close to two, diamonds are spread almost uniformly throughout the world, if it is close to zero, there are a few privileged places where all of the diamonds are concentrated. In the first situation we soon find something, but the yield is very low; in the second case we must search for a long time but then we will be well paid for our efforts. We quickly realize that the information provided by the fractal dimension is inadequate. There is a *measure attached to the support of this fractal set*, which is the price of diamonds as a function of their volume, clearly it is more worthwhile to find large stones than small ones. We must use our knowledge of the distribution of diamonds by bringing in the parameter of their size. Let us suppose that we know the distribution perfectly (otherwise we must take a sample) and let us cover the globe with a grid, attaching to each of its square plots of side ε the monetary value of the diamonds found there. To simplify matters the set of plots of land can be divided into a finite number of batches ($i = 1,...,N$)

corresponding to the various slices of value, from the poorest to the richest [to each plot i we attach in this way its value $\mu(\varepsilon,x_i)$ relative to the total value]. The correspondence of each of these batches to a given slice μ specifies a distribution of diamonds on the earth's surface; we assume that each of these distributions is fractal in the limit when the side ε of the plots tends to zero. The very rare, rich regions will have a fractal dimension close to zero ("dust"), whereas the regions with only a trace of diamonds, albeit uniformly distributed, will have a fractal dimension close to two.

The multifractal character is connected with the heterogeneity of the distribution (see Sec. 1.4.2, and T.A. Witten in Les Houches, 1987). For a homogeneous fractal distribution, the mass in the neighborhood of any point in the distribution is arranged in the same manner. That is to say that inside a sphere of radius R centered on the fractal at x_i, the mass $\mathcal{M}(R)$ fluctuates little about its mean value over all the x_i, $\langle\mathcal{M}(R)\rangle$ whose scaling law is R^D: the distribution $P(\mathcal{M})$ of the masses $\mathcal{M}(R)$ taken at different x_i is narrow, that is, it decreases on either side of the mean value faster than any power. In particular, all the moments vary like $\langle\mathcal{M}(R)^q\rangle \propto \langle\mathcal{M}(R)\rangle^q$, for all q. The fractal distribution is described by the sole exponent D. This is not so for heterogeneous fractals for which there is a broad distribution, $P(\mathcal{M})$, of the masses. Such is the case for the distribution of diamonds on the earth's surface. Knowledge of the behavior of the moments $\langle\mathcal{M}(R)^q\rangle$ tells us about the edges of the distribution, namely the very poor and the very rich regions.

We are going to make these ideas sharper using some simple distributions which will allow us to introduce some mathematical relationships indispensable in the practical use of the concept of multifractility.

The quantities $f(\alpha)$ and $\tau(q)$ (which we are now going to define) will allow us to characterize the distributional heterogeneities of the measures known as *multifractal measures*. As these ideas are not initially obvious, the reader may, if he wishes, turn to the various examples given further on.

1.6.1 Binomial fractal measure

This simple measure is constructed as follows: a segment of length L, on which a uniform measure of density 1/L is distributed, is divided into two parts of equal length: $\ell_0 = \ell_1 = L/2$ to which the weights p_0 and p_1 are given (p_0 to the left and p_1 to the right) (Fig. 1.6.1). This process is iterated ad infinitum. The total measure is preserved if care is taken in choosing

$$p_0 + p_1 = 1.$$

Each element (segment) of the set is labeled by the successive choices (0 left, 1 right) at each iteration. At the nth iteration, each segment is thus indexed by a sequence $[\eta] \equiv [\eta_0, \eta_1, ..., \eta_k,... \eta_n]$ where $\eta_k = 0$ or 1, and has length, $d\ell = \varepsilon L = 2^{-n} L$. Its abscissa on the segment E=[0,1] is described simply by the number in base two

$$x = \ell / L = 0. \eta_0\eta_1...\eta_n.$$

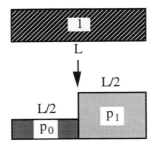

Fig 1.6.1. First iteration of the binomial measure. The measures associated with the areas of the rectangles are shown inside them.

So, at the third iteration, the successive weights $\mu(\epsilon,x_i)$ are the $p_{[\eta]}$ where

$[\eta] = [000]$ $(x_0=0.000)$ $\rightarrow$ $p_{[000]} = p_0^3$

     $[001]$ $(x_1=0.001)$

     $[010]$ $(x_2=0.010)$ $\rightarrow$ $p_{[001]} = p_{[010]} = p_{[100]} = p_0^2 p_1$

     $[100]$ $(x_3=0.100)$

     $[011]$ $(x_4=0.011)$

     $[101]$ $(x_5=0.101)$ $\rightarrow$ $p_{[011]} = p_{[101]} = p_{[110]} = p_0 p_1^2$

     $[110]$ $(x_6=0.110)$

     $[111]$ $(x_7=0.111)$ $\rightarrow$ $p_{[111]} = p_1^3$

and so on for each value of n (or of $\epsilon = 2^{-n}$). For each n, the distribution is normalized to one:

$$\sum_i \mu(\epsilon,x_i) = \sum_{[\eta]=[00...0]}^{[11...1]} p_{[\eta]} = 1 .$$

It can easily be seen that the weight associated with a sequence $[\eta]$ has the general form

$$\mu(\epsilon,x) \equiv p_{[\eta]} = p_0^{n\varphi_0} p_1^{n\varphi_1}$$

[$n\varphi_0$ and $n\varphi_1$ being the number of 0's and 1's in $[\eta]$ respectively: $\varphi_0 = k/n$ and $\varphi_1 = (n-k)/n$, $k = 0,...n-1$].

To each value of x is associated a $\varphi_0(x) = 1-\varphi_1(x)$. These weights occur with a frequency

$$N(\epsilon,x) = \binom{n}{k} = \frac{n!}{(n\varphi_0)!(n\varphi_1)!} .$$

Fig. 1.6.2 shows the hierarchy of iterations for $n = 4$, and the corresponding distribution of weights (more precisely their logarithm). The logarithm of the weight on an interval ϵ, divided by the logarithm of that interval, is called the *Holder exponent* and is denoted by α ,

$$\alpha = \frac{\log \mu(\varepsilon)}{\log \varepsilon} = -\varphi_0 \log_2 p_0 - \varphi_1 \log_2 p_1$$

It measures the *singular behavior of the measure* in the neighborhood of a point x [via $\varphi_0(x)$ and $\varphi_1(x)$], thus

$$\boxed{\mu(\varepsilon,x) = \varepsilon^{\alpha(x)}}.$$

(1.6-1)

The values of α are bounded according to the following inequality:

$$0 < \alpha_{min} = -\log_2 p_0 \le \alpha \le \alpha_{max} = -\log_2 p_1 < \infty.$$

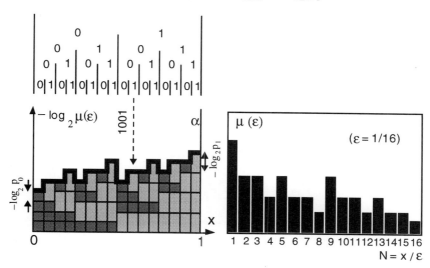

Fig. 1.6.2. (Left) the binary sequences leading to the [η] for n=4. The graph on the right shows the distribution of the measures μ whose logarithm is also proportional (factor –n) to α. Here we have chosen log p_1 = 2 log p_0. We can see the symmetry of the binomial distribution: contrary to the example of the diamonds, the regions of low weight are as rare as those with high weight.

This behavior allows us to partition[8] the set E into subsets having the same α,

$$E = \bigcup_{\alpha} E_\alpha .$$

(1.6-2)

Let us now take the subsets E_α into closer consideration. We enumerated them above while calculating N(ε). The analog of the Hausdorff dimension for the support of intervals with the same α (function of α and therefore of x) is then written

---

[8] In our intuitive example about the distribution of diamonds the partition into "batches" is made according to their relative "value;" we see here that these batches may also be designated by their Holder exponents which indicates how the value of a piece of land varies (locally) as a function of its size.

$$\delta(\alpha) = -\frac{\log N(\epsilon,\alpha)}{\log \epsilon} \quad . \tag{1.6-3}$$

In the limit of large n (small ϵ) we have simply

$$\delta = -\log [(n\varphi_0)!(n\varphi_1)!/n!] / \log \epsilon$$
$$\approx -\varphi_0 \log_2 \varphi_0 - \varphi_1 \log_2 \varphi_1 \quad .$$

In the binomial case, this fractal dimension is well defined for each set E_α since all the parameters are determined [via $\varphi_0(x)$ and $\varphi_1(x)$].

1.6.2 Multinomial fractal measure

The results just shown for a *binomial measure* easily generalize to *multinomial measures*. By considering b weights p_β $(0 \le \beta \le b-1)$, we can go through an analogous procedure to that of the binomial measure. Each segment of size b^{-n} at iteration n is indexed by a sequence $[\eta]$ or an abscissa x, written in base b (instead of the 0's and 1's of the previous binomial example). The b-adic intervals are then characterized by the frequencies φ_β of their "digits" in base b. So, the expressions for α and δ generalize to

$$\alpha = -\sum_\beta \varphi_\beta \log_b p_\beta \quad \text{and} \quad \delta = -\sum_\beta \varphi_\beta \log_b \varphi_\beta$$

with the constraints $\sum_\beta \varphi_\beta = 1$ and $\sum_\beta p_\beta = 1$.

In the case of a binomial measure (b = 2), δ is a single valued function of α, since two parameters (φ_0 and φ_1), whose sum is normalized to one, are used. For b > 2, this relationship is no longer single valued (there are b–2 supplementary parameters) and the pairs (α,δ) cover a certain domain. This domain is roughly indicated by a network of curves in Fig. 1.6.3 (for b = 3). The set of $[\eta]$ (or x expressed in base b) corresponding to the same α is dominated by the term of highest dimension, f = max δ [i.e., by the subset $N(\epsilon,\alpha)$ whose exponent δ is the greatest]:

$$N(\epsilon,\alpha)_{\text{dominant}} \propto \epsilon^{-f(\alpha)} \quad . \tag{1.6-4}$$

This term therefore maximizes

$$-\sum_\beta \varphi_\beta \log_b \varphi_\beta \quad .$$

The variation is thus written, (the $\vartheta\varphi_\beta$ then being independent infinitesimal variables),

$$\vartheta \left(-\sum_\beta \varphi_\beta \log_b \varphi_\beta\right) \equiv 0$$

with the constraints

$$\alpha = -\sum_{\beta} \varphi_{\beta} \log_b p_{\beta} \quad \text{and} \quad 1 = \sum_{\beta} \varphi_{\beta} \ .$$

This is solved in the classical way by introducing Lagrange multipliers q and (r–1), leading to the relation

$$\sum_{\beta} \vartheta \varphi_{\beta} [\log_b \varphi_{\beta} - q \log_b p_{\beta} - r] \equiv 0 \qquad \forall \ \vartheta \varphi_{\beta}.$$

All the terms inside the brackets vanish. Hence, since the φ_{β} are normalized to unity (thus determining r):

$$\varphi_{\beta} = \varphi_{\beta} (q) = b^r \ p_{\beta}^q = \frac{p_{\beta}^q}{\sum_{\beta} p_{\beta}^q} \ , \qquad -\infty < q < +\infty.$$

The φ's dependence on x now operates via q which picks out the subsets $\{x\}(q)$ corresponding to $\delta(\alpha)$ extreme for fixed α.

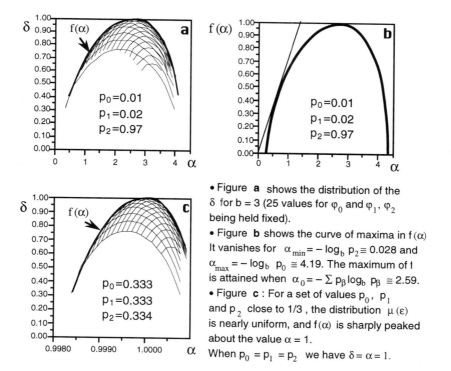

• Figure **a** shows the distribution of the δ for b = 3 (25 values for φ_0 and φ_1, φ_2 being held fixed).

• Figure **b** shows the curve of maxima in $f(\alpha)$ It vanishes for $\alpha_{min} = -\log_b p_2 \cong 0.028$ and $\alpha_{max} = -\log_b p_0 \cong 4.19$. The maximum of f is attained when $\alpha_0 = -\sum p_{\beta} \log_b p_{\beta} \cong 2.59$.

• Figure **c** : For a set of values p_0, p_1 and p_2 close to 1/3 , the distribution $\mu(\epsilon)$ is nearly uniform, and $f(\alpha)$ is sharply peaked about the value $\alpha = 1$.

When $p_0 = p_1 = p_2$ we have $\delta = \alpha = 1$.

Fig. 1.6.3. Distribution of the Hausdorff dimensions of subsets E_{α}.

It is then useful to introduce the quantity

$$\tau(q) = \log_b \sum_\beta p_\beta^q .$$
(1.6-5a)

Very generally $\tau(q)$ enters in the framework of cumulant generating functions. Note that some authors have adopted the opposite sign.

α and $f(\alpha)$ are simply related to $\tau(q)$ by the following relations:

$$\alpha = -\sum_\beta \varphi_\beta \log_b p_\beta = -\frac{\partial}{\partial q} \log_b \sum_\beta p_\beta^q$$

and $\max \delta = f(\alpha) = -\left(\sum_\beta p_\beta^q [q \log_b p_\beta - \log_b \sum_\beta p_\beta^q] \right) / \sum_\beta p_\beta^q$

which reduce to the remarkable equations

$$\alpha = -\frac{\partial \tau(q)}{\partial q} \quad \text{and} \quad f(\alpha) = \tau(q) - q\frac{\partial \tau(q)}{\partial q} = \tau + q\alpha$$
(1.6-5b)

with $\alpha = -\sum_\beta \varphi_\beta \log_b p_\beta$, $f(\alpha) = -\sum_\beta \varphi_\beta \log_b \varphi_\beta$,

$$\varphi_\beta = \frac{(p_\beta)^q}{\sum_\beta (p_\beta)^q} \quad \text{and} \quad \tau(q) = \log_b \sum_\beta (p_\beta)^q, \quad -\infty < q < +\infty .$$
(1.6-5c)

These results (constituting the formalism of multifractals) were obtained by Frisch and Parisi (1985) and Halsey *et al.* (1986) using the method of steepest descent. The concept itself already existed in a 1974 paper of B. Mandelbrot. Here we have followed the more straightforward approach of B. Mandelbrot (1988).

It can be seen that the functions $f(\alpha)$ and $\tau(q)$ are Legendre transforms of each other (Fig. 1.6.4). Such transforms are frequently used in thermodynamics when one wishes to change independent variables.

Formally, f may be compared with an entropy, q with the reciprocal of a temperature, α with an energy (conjugate variable of q), and τ with a free Gibbs energy (Lee and Stanley, 1988). The expression for f is indeed that of an entropy or, more precisely, that of a quantity of information, provided that the φ_β are considered to be the probabilities of finding the measures p_β (the relative frequences of the digit β in $x = 0.\eta_0\eta_1...\eta_n$ written in base b). The minimum entropy is obtained when the distribution of the p_β is known exactly, that is, when one of the $\varphi_\beta = 1$ and the others zero (e.g., $x = 0.333...33$). On the other hand, the maximum disorder or entropy corresponds to all the φ_β being equal [here the φ_β equal $1/b$ and $f(\alpha) = 1$]. The expression for $\varphi_\beta(q)$ shows that this maximum disorder corresponds either to $q = 0$, or to the trivial case $p_0 = \ldots p_\beta = \ldots p_{b-1} (= 1/b)$.

Generally speaking, $f(\alpha)$ is a convex, positive curve, increasing from $f(\alpha_{min}) = 0$ to a maximum $D_0 = \max f(\alpha)$, which is equal to one in the previous example; then $f(\alpha)$ decreases to $f(\alpha_{max}) = 0$. The extreme values of α are given simply by the conditions of minimum entropy (one of the $\varphi_\beta = 1$, all the rest $= 0$):

$$\alpha_{min} = \min (-\log_b p_\beta), \quad \alpha_{max} = \max (-\log_b p_\beta).$$

Significance of $\tau(q)$

Consider the following measure:

$$M_q(\varepsilon) = \sum_i \mu(\varepsilon, x_i)^q . \tag{1.6-6}$$

What does this measure represent?

When $q = 0$, $M_0(\varepsilon)$ represents the volume of support measured in intervals ε, $M_0(\varepsilon) \approx (L/\varepsilon)^{D_0}$, with $D_0 = 1$ here, for the segment $[0,1]$ of length $L = 1$.

When $q = 1$, $M_1(\varepsilon)$ represents the sum over the support of the measures on the ε intervals. As this measure is normalized,

$$M_1(\varepsilon) \equiv \sum_{[\eta]=[00...0]}^{[11...1]} p_{[\eta]} = 1.$$

When $q \to +\infty$, M_q is dominated by the regions of high density μ/ε.
When $q \to -\infty$, M_q is dominated by the regions of low density μ/ε.

Thus, the parameter q allows us to select subsets E_q corresponding to higher or lower densities.

We usually put

$$\boxed{M_q(\varepsilon) = \varepsilon^{(q-1)D_q}} ; \tag{1.6-7}$$

an expression which is true for $q = 0$ and $q = 1$. D_q are called *qth order generalized dimensions* [9] (the name is due to Hentschel and Procaccia, 1983), although they are only a dimension when $q = 0$ (D_q can however be defined as a critical dimension when $q > 1$). The advantage of D_q over $\tau(q)$ resides essentially in the fact that the former all reduce to the fractal dimension D when the space is homogeneous, that is,

$$\mu(\varepsilon, x) \propto \varepsilon^D \quad \forall x,$$

for then

$$M_q(\varepsilon) = \sum_{support} \mu(\varepsilon, x)^q \propto \sum_{support} \varepsilon^{qD} \propto \varepsilon^{-D} \varepsilon^{qD} , \text{ and hence } D_q = D.$$

Finally, it can be shown that D_q decreases monotonically as q increases.

---

[9] Sometimes also called *Renyi dimensions*.

We shall now show that $(1 - q)D_q = \tau(q)$. From above,

$$M_q(\varepsilon) = \sum_i \varepsilon^{q\,\alpha(x_i)} \quad .$$

Now the number of domains corresponding to the same α is known, it is $N(\varepsilon,\alpha) \propto \varepsilon^{-\delta(\alpha)}$, and hence

$$M_q(\varepsilon) \approx \int_{\alpha_{min}}^{\alpha_{max}} d\alpha\; \varepsilon^{q\,\alpha - \delta(\alpha)}.$$

This integral is dominated by the maxima of $\delta(\alpha)$ with the value of α which minimizes the exponent $(\varepsilon \ll 1)$, that is, $\alpha(q)$ such that

$$\frac{\partial}{\partial \alpha}\, [q\,\alpha - \max \delta(\alpha)]\Big|_{\alpha=\alpha(q)} = 0, \quad \text{where} \quad \max \delta(\alpha)\Big|_{\alpha=\alpha(q)} \equiv f(\alpha(q))$$

which agrees with the earlier results. By comparison it can be seen that $M_q(\varepsilon) \propto \varepsilon^{q\,\alpha - f}$, hence

$$M_q(\varepsilon) \propto \varepsilon^{-\tau(q)}$$

$$\text{with} \quad \tau(q) = (1-q)\,D_q = -\lim_{\varepsilon\to 0} \frac{1}{\log \varepsilon}\, \log \sum_i \mu(\varepsilon,x_i)^q \quad . \qquad (1.6\text{-}8)$$

(The integral is a discrete sum if the box method is used.)

Form and meaning of D_1

When $q = 1$, the above expression is undetermined. So what is the form and the meaning of D_1, the first order generalized dimension? For this we must calculate

$$D_1 = \lim_{q\to 1} \frac{1}{q-1}\,\lim_{\varepsilon\to 0} \frac{\log M_q(\varepsilon)}{\log \varepsilon}. \qquad (1.6\text{-}9)$$

Writing $[\mu(\varepsilon,x)^q] = (\mu\,\mu^{q-1}) \cong \mu[1+ (q-1)\log \mu]$, so that

$$\log M_q(\varepsilon) = \log \sum_i \mu(\varepsilon,x_i)^q \cong \log\left[1+ \sum_i (q-1)\mu(\varepsilon,x_i)\log \mu(\varepsilon,x_i)\right]$$

$$\cong (q-1)\sum_i \mu(\varepsilon,x_i)\log \mu(\varepsilon,x_i)$$

gives

$$D_1 = \lim_{\varepsilon\to 0} \frac{1}{\log \varepsilon}\sum_i \mu(\varepsilon,x_i)\log \mu(\varepsilon,x_i) \quad . \qquad (1.6\text{-}10)$$

Moreover, we can relate this expression to α and $f(\alpha)$ by differentiating $\tau(q)$. We then find that

$$D_1 = \alpha_{q=1} = f(\alpha_{q=1}). \qquad (1.6\text{-}11)$$

The remarkable property of D_1 comes from the fact that $\sum \mu(\varepsilon)\log \mu(\varepsilon)$ represents the entropy of information of the distribution whose scale behavior

D_1 describes. For this reason D_1 is called the *information dimension*. The support set $E_{\alpha(1)}$ contains *almost all the measure* (or mass) of the set E.

In particular, for the multinomial measure on [0,1], we have $\varphi_\beta = p_\beta$ (as q = 1) which means that $D_1 = \Sigma\, p_\beta \log_b p_\beta$. The *information dimension*, as we showed above, reaches a maximum when all the p_β are equal (to 1/b), the information about the distribution then being minimal. On the other hand, D_1 is zero when all the p_β are zero except one, p_β. The information is then complete since the whole measure is on one abscissa point x = 0. $\beta\beta\beta$...

To finish, let us now calculate the mass exponent of the measure distribution $\mu(\varepsilon,x)^q$ using the η-dimensional Hausdorff measure (see Sec. 1.3). It is found as the sum of volume elements ε^η, weighted by their associated measure μ^q, that is,

$$m_\eta\,(E,\mu^q) = \lim_{\varepsilon\to 0}\, \sum_i \mu(\varepsilon,x_i)^q \varepsilon^\eta = \lim_{\varepsilon\to 0}\, M_q(\varepsilon)\, \varepsilon^\eta\,.$$

From what we saw earlier we find that

$$m_\eta\,(E,\mu^q) = \begin{cases} \to 0 & \text{if } \eta > \tau(q) \\ \to \infty & \text{if } \eta < \tau(q) \end{cases}.$$

$\tau(q)$ is therefore the mass exponent of the distribution μ^q. This result is trivial here because there is only one dilation scale (a factor 1/2 for each segment at each iteration). When *nonuniform fractals* are considered, several dilation scales are present and this formulation proves to be the most direct method of calculating $\tau(q)$ and the other multifractal characteristics.

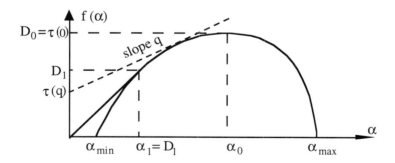

Fig. 1.6.4. *The Legendre transform allows one to pass from the equation f = f(α) of a plane curve to the equation τ = τ(q) of the same curve by eliminating α between q = df/dα and τ = f – qα.*

Significance of the maximum of f(α)

The maximum of f(α) corresponds to $\partial f/\partial\alpha \equiv q = 0$, and consequently to all the φ_β equaling 1/b,

$$M_0(\varepsilon) = (L/\varepsilon)^{D_0} \quad \text{with} \quad \tau(0) = \max f(\alpha) = f(\alpha(0)) = D_0.$$

$D_0 = 1$ in the multinomial example.

The maximum of $f(\alpha)$ therefore corresponds to a uniform measure on the support. Its value is thus the dimension (fractal or otherwise) of the support. Here, it is that of a segment $[0, L]$ and therefore equals one.

Everything that has been developed for the multinomial case will contribute to understanding the general case of a distribution of measures over a fractal support. We shall approach this subject by explaining in detail two important measures for physics, the multifractal measure of a distribution of points, mass or current (developed turbulence, strange attractors, distributions of galaxies or flux of matter in a porous rock, resistors networks), and the harmonic multifractal measure (growth phenomena, DLA, etc.). Remaining within the general framework, some structures which are fractal on several scales, with multifractal characteristics which may be fully calculated, will be considered.

1.6.3   Two-scale Cantor sets

The Cantor sets mentioned in Sec. 1.4.1 with their uniform measures are not multifractal; in fact their curves $f(\alpha)$ reduce to a point $[\alpha = f(\alpha) = D]$. They are pure fractals. To generate a multifractal structure with a uniform measure at least two dilation scales must be used.

Let us examine one of these structures by treating the general case: two measure scales, two dilation scales. The following structure (Cantor bars) generalizes the binomial measure studied above. An initial segment of length L and unit measure (mass) is divided into two parts: ℓ_0 and ℓ_1 to which we associate the measures p_0 and p_1 ($p_0 + p_1 = 1$). The first three steps of the construction are shown in Fig. 1.6.5.

Determination of $\tau(q)$

$\tau(q)$ will be determined by calculating the η-dimensional Hausdorff measure. In practice the result is completely analogous to that giving the fractal dimension calculated from Eq. 1.4-2. We therefore calculate

$$m_\eta(q,N) = \sum_{i=0}^{N-1} \mu_i^q \, \ell_i^\eta = \underset{\varepsilon\to 0}{\longrightarrow} \begin{cases} \to 0 \ \text{ if } \eta > \tau(q) \\ \to \infty \ \text{ if } \eta < \tau(q) \end{cases} \quad \text{where } \varepsilon = \max(\ell_i).$$

In the present case, $i = [0, N-1]$ designates the i th element out of $N = 2^n$, (n being the number of iterations). It takes the place of x in the binomial measure. The measure and the size of the i th element are, respectively,

$$\mu_i = p_0^k \, p_1^{n-k} \quad \text{and} \quad \ell_i = \ell_0^k \, \ell_1^{n-k}$$

and its degeneracy is the number of ways of choosing k objects out of n without regard to order of choice. Thus,

$$m_\eta\,(q,N\,) = \sum_{k=0}^{n}\,\binom{n}{k}\,(p_0^k\,p_1^{n-k})^q\,(\,\ell_0^k\,\ell_1^{n-k})^\eta = \Big(\,p_0^q\,\ell_0^\eta + p_1^q\,\ell_1^\eta\Big)^n$$

$$m_\eta\,(q,N\,) \xrightarrow[n\to\infty]{}\begin{cases} \to 0 \ \ \text{if}\ \ p_0^q\,\ell_0^\eta + p_1^q\,\ell_1^\eta <1 \ \ (\,\eta > \tau(q)\,) \\ \to \infty \ \ \text{if}\ \ p_0^q\,\ell_0^\eta + p_1^q\,\ell_1^\eta >1 \ \ (\,\eta < \tau(q)\,) \end{cases}$$

and $\tau(q)$ is determined by the equation [analogous to that of Sec.1.4.1, Eq. (1.4-2)]

$$p_0^q\,\ell_0^{\tau(q)} + p_1^q\,\ell_1^{\tau(q)} = 1$$

Knowing $\tau(q)$ we can also calculate D_q, α, and $f(\alpha)$. For a generalization to several scales see Hentschel and Procaccia (1983).

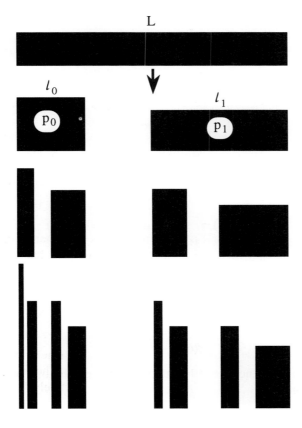

Fig. 1.6.5. The three first iterations of the Cantor set weighted by dilation scales $\ell_0 = L/4$ and $\ell_1 = L/2$ and weights $p_0 = 0.4$, $p_1 = 0.6$.

1.6.4   Multifractal measure on a set of points

Any statistical description of a set of points uses the notion of correlation, more or less directly. Indeed, these correlations represent deviations from a uniform distribution of points (i.e., translational invariance).

One way of determining these correlations is by calculating the moments of the distribution of points. For this the box-counting method may be used (see Sec. 1.3.3) and a measure defined in each box of side ε centered at x.

The number of points inside a box is $\mathcal{N}(\varepsilon,x)$, so that the probability of finding a point is $\mathcal{N}(\varepsilon,x)/\mathcal{N}$, $\mathcal{N}$ being the number of points in the set.

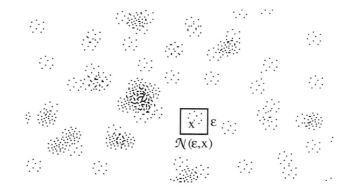

Fig. 1.6.6. Multifractal measure of a cloud of points. Use of the box-counting method.

$$\mu(\varepsilon,x) = \mathcal{N}(\varepsilon,x)/\mathcal{N}$$

will therefore represent our measure. Its $(q-1)$th order moment may be calculated by summing over the boxes:

$$\langle \mu(\varepsilon)^{q-1} \rangle = \sum_{i \in boxes} \mu(\varepsilon,x_i)^q = M_q(\varepsilon) .$$

If the structure is multifractal, from above, μ and M display the power laws,

$$\mu(\varepsilon,x) \propto \varepsilon^{\alpha(x)} \quad \text{and} \quad M_q(\varepsilon) \propto \varepsilon^{-\tau(q)} = \varepsilon^{(q-1)\,D_q}$$

the exponent $\alpha = \alpha(q)$ is the same for all $x \in E_{\alpha(q)}$.

Having found $\tau(q)$ via

$$\tau(q) = -\frac{\log M_q(\varepsilon)}{\log \varepsilon}$$

$\alpha(q)$ and $f(\alpha)$ may then be calculated, using the relations obtained for the multinomial measure

$$\alpha = -\frac{\partial \tau(q)}{\partial q} \quad \text{and} \quad f(\alpha) = \tau + q\alpha,$$

since their derivation is general. What is new here, in relation to the multinomial measure, is that the support (set of points) may itself be fractal. This description may be used to characterize the strange attractors obtained in chaotic phenomena. We shall give some examples of this in Sec. 2.3.

For a more complete (but also more difficult) justification of the above formalism, Collet et al. (1987), Bohr and Rand (1987), Rand (1989), and Ruelle (1982, 1989) may be consulted.

CHAPTER 2

Natural Fractal Structures
From the macroscopic...

Nature provides numerous examples of fractal structures ranging from those with a fraction of the scale of the universe down to those on the atomic scale. These structures can be observed in the distribution of galaxies, cloud structures, mountain reliefs, turbulent flows on the surface of a planet like Jupiter, fractured rocks, rough surfaces, disordered materials, aggregates of particles and atoms, etc. We shall start with a description of "giant" fractal structures. The existence of such large structures should come as no surprise as the very essence of the notion of fractals and critical phenomena is involved here: For a physical process generating fractal structures in which, by construction, there is no internal scale, there is no reason for there to be any upper size limit to these structures, so long as the parameters of the process are conveniently chosen. This is what we are going to show in the rest of this chapter.

We shall take advantage of various examples described here to introduce new mathematical ideas and physical models. Although the theme of this chapter concerns giant fractal structures, we shall also mention, as we go along, fractal structures possessing similar characteristics but on totally different scales.

2.1 Distribution of galaxies

From the first decades of the 20th century, astronomers have noticed the hierarchical "clustering" of galaxies. Recent observational results show a universe which is highly structured, yet equally highly disordered. It turns out to be less simple than was imagined: on a scale much greater than that of galactic clusters, matter seems to be arranged in flattened, stringlike, and holey structures. Progress in the analysis of clusters is closely related to theoretical progress in the study of the origin of cosmic structures, and it is important to compare the various "Big Bang" models predicting the distributions of galaxies with the observations.

2.1.1 Distribution of clusters in the universe

Galaxies are found to be distributed in groups on all scales. Our galaxy, the "Milky Way", belongs to what is called the "local group," comprising approximately 20 galaxies; its size is around[1] 1 Mpc. The nearest cluster is that of Virgo at around 10 Mpc from us; then there is Coma (Berenice's hair): formed of several thousand galaxies at 100 Mpc. These clusters themselves make up a supercluster. We belong to the local supercluster, discovered by de Vaucouleurs around 1958. It has the shape of a disk approximately 1 Mpc wide and 20 Mpc long. This local hierarchy to which our galaxies belong is actually a very common structure.

From observations the following distributions of masses and distances may be ascertained :

Galaxies:
> Their mass represents $\approx 10^7$ to $10^{12} M_\odot$ (solar masses)
> and their size $\approx$ 10 to 100 Kpc.

Half the galaxies belong to groups of several tens of galaxies with
> mass $\approx 10^{12}$ to $10^{14} M_\odot$
> and size $\approx$ 1/10 to 10 Mpc.

Clusters are formed of several thousand galaxies with
> mass $\approx 10^{15} M_\odot$
> and size $\approx$ 10 Mpc.

The essential difficulty in three-dimensional cartography of the universe is the problem of distances. The distance of the majority of galaxies has been measured using their redshift z: in the conventional interpretation the expansion speed v and z are related to the distance D of the object by Hubble's law $v = cz = H_0 D$. Two difficulties then appear: the constant H_0 is not known accurately (between 50 and 100 km/s per Mpc, 82 seems a good recent value) and, superimposed on the expansion, galaxies have their own unknown speeds. Other methods of measuring these distances exist, but they use up too much computer time and distances have only been measured in this way for a few galaxies. This explains the controversies over the extent of the fractal domain.

The theory of star and galaxy formation, due to Hoyle, the descriptive model of Fournier d'Albe, and above all the empirical data unanimously suggest a large zone of internal similarity in which the fractal dimension is close to $D = 1$. But let us return to the last century and to the questions which were asked then.

---

[1] The usual units of distance and mass in this field are the *parsec* (pc) and the *solar mass* ($M_\odot$), respectively. A parsec is the distance at which an astronomical unit (i.e., the mean distance between the earth and the sun, 1 U.A.$\approx$150 million km) subtends an angle of one second of an arc, 1 pc $\approx$ 3.08 x 10^{13} km (a light year corresponds to 9.45 x 10^{12} km, i.e., about 0.307 pc).

2.1.2 Olbers' blazing sky paradox

Olbers' paradox (1823) consists in noticing that if the distribution of celestial bodies were uniform then the night sky would not be dark. The luminosity density of a star is (roughly) the same for all stars of the same type, for, if it is situated at a distance R, its luminosity decreases as $1/R^2$ but this is also true for its apparent diameter. With a uniform distribution throughout the universe, almost all directions issuing from a terrestrial observer cut the apparent disk of a star, at least if the universe is infinite.[2] It has also been shown that even for a realistic finite universe the luminosity of the sky would be several orders of magnitude too high. The paradox disappears if $\mathcal{M}(R) \propto R^D$ with $D < 2$, for then a significant proportion of directions do not intersect any star[3] (see Figs 2.1.1 and 2.1.2). Indeed the hypothesis of a fractal distribution is sufficient but not necessary.[2]

Fournier d'Albe expressed this idea at the beginning of the century in a totally unrealistic model. Below we have represented Fournier's very simple hierarchical model. (The scale and mass factors have been chosen to equal 5 giving D = 1.) With such a distribution the sky would appear sprinkled with stars on a black background (Fig. 2.1.1):

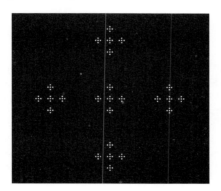

Fig. 2.1.1. Hierarchical model of Fournier d'Albe. Here the fractal dimension is D = 1.

Around 1908 Charlier constructed a hierarchical model which was more random at each level of the hierarchy and therefore more realistic than Fournier's nonstatistical hierarchical model (1907). (The latter did however already contain the interesting idea). Then to introduce a notion of fractal structure into the mass distribution, Mandelbrot proposed (Mandelbrot, 1982,

---

[2] Olbers' paradox is in fact resolved if one accepts the hypothesis of the Big Bang and the expanding universe, since the domain observable from Earth is then finite (objects at a distance greater than λct cannot be seen, c being the speed of light and t the age of the universe, $\lambda < 1$).

[3] For separate reasons we must have $D < 3/2$.

p. 294) the model of a "scattered" universe.

Mandelbrot's scattered universe, of dimension D, is simulated by a *Levy flight*: let S_i and S_j (S for star) be two successive points where objects are situated. They are chosen such that S_iS_j has an arbitrary direction and the length $|S_iS_j| = U$ is distributed according to the probability law Prob (U>u) = u^{-D} when u >1, and Prob (U>u) = 1 when u < 1; D is found to be the fractal dimension. To satisfy the condition $\langle U^2 \rangle = \infty$ (the distribution is not bounded), we need 0 < D < 2.

A collection of objects is thus obtained for which there is a broad distribution in the distances between them, with the probability distribution function slowly decreasing as the distance increases. Schematically, the objects have a tendency to group into a hierarchy (see Fig. 2.1.2).

Random directions issuing from an Earthbound observer E almost certainly have a zero probability of intersecting a stellar disk S_i; the observer therefore sees an essentially dark sky.

Representation by a Levy flight is intuitive (or *ad hoc*) and not based on the physics of star and galaxy formation. However, it does provide a fairly convincing representation of the manner in which celestial bodies must be distributed to be in agreement with the observations described above.

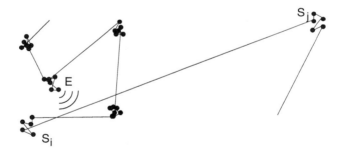

Fig. 2.1.2. *Path of a Levy flight, random walk whose mean free path diverges when D < 1. The arrival points of the jumps simulate a distribution of celestial objects of dimension D. The objects (stars) are represented by black points in the drawing, and the lines show the path taken during the construction.*

But this statistical structure is still too simplistic [which is why Mandelbrot has also proposed a dieresis model (Mandelbrot, 1982)]. From tables, despite inaccuracies over the distances, the following result has been found: the universe is formed of groups up to 20 Mpc across with a fractal dimension: D $\cong$ 1.23 $\pm$ 0.04, over a domain of the order of 20 Mpc [4] (de Vaucouleurs, 1970; Pietronero, 1987; Coleman et al., 1988; Peebles, 1989).

---

[4] Opinions still vary about this point: Peebles believes the fractal dimension to extend to 5 Mpc, while Pietronero and Mandelbrot believe it extends indefinitely.

$$\Gamma(r) = \frac{\langle n(\vec{r}_0)\, n(\vec{r}_0 + \vec{r})\rangle}{\langle n(\vec{r}_0)\rangle}$$

where $n(\vec{r})$ is the local mass density in $\vec{r}$. $\langle n \rangle$ is its mean value, dependent on the sample size, while $\Gamma(r)$ does not depend on it (see Pietronero, 1987; Blanchard and Alimi, 1988; Coleman in Pietronero, 1989 for a discussion of this point). Another difficulty is that the catalogs do not provide the galaxies' masses, only their positions (i.e., their directions and red shift). Moreover the distribution of dark matter is not well known. The structure seems to be different over greater distances (>20 Mpc) and the field remains largely open and controversial. It must be kept in mind that the distribution statistics are made on earth at a particular "moment" and that an instantaneous photographic image of objects in space does not correspond to "reality", relativistic effects being inevitable on these scales.

Let us end this topic of mass distributions in the universe by remarking that in 1895, Seeliger, from the knowledge that gravitational forces also decrease as R^{-2}, sought, preceding Charlier's ideas, conditions for which the gravitational force and Newtonian potential did not diverge. In 1930, Paul Lévy (Lévy, 1930) independently took this problem up again, and he too suggested a hierarchical model to satisfy this constraint. This model was a fractal structure! Unfortunately, the problem of gravitational forces is more complicated than that of starlight, since these forces partially compensate one another. Finally, we note that Balian and Schaeffer have proposed a dynamical explanation of the scaling laws of galaxies (Balian and Schaeffer, 1989).

2.2 Reliefs, clouds, fractures...

Mandelbrot noticed early on that if coastlines display fractal characteristics, it must be equally possible to generate various natural objects, such as mountain reliefs, craters on the moon, or clouds, numerically. (Computer generated landscapes used in the production of certain science fiction films are certainly some of the costliest applications of fractals !). The simplest method of creating an artificial relief is based on the Brownian function of a point, defined by P. Lévy (1948), or more precisely on fractional Brownian functions, which we shall define below.

Let us point out immediately that it is essentially a visual approach and that the conceptual understanding of relief structures is closely related to fracture propagation processes, a subject still imperfectly understood.

We are going to take the shapes of coastlines as a pretext for introducing considerations about *Brownian motion* and its generalization, *fractional Brownian motion*. Random walks, at the root of Brownian motion, will also be used in connection with the structures of polymer chains.

Moreover, recordings of the variation in time of various natural parameters such as temperature, rainfall, rate of flow of rivers, or even stock exchange prices, display structures which can be represented by fractional Brownian motion. The first empirical laws concerning the movements of these natural phenomena are due to Hurst. Hurst was particularly interested in variations in reservoir and river levels (Hurst, 1951; Hurst et al., 1965), for which he devised a method (R/S analysis) allowing these variations to be characterized by means of an exponent, since named the Hurst exponent, H.

2.2.1   Brownian motion and its fractal dimension

A Brownian motion is represented by a series of jumps r_i, randomly directed and of equal length or possibly of length U, itself random but having a finite mean value a. The latter case corresponds to an example of what is called a *Rayleigh flight*, which has the property that

$$\text{Prob } (U>u) = \exp (-u\sqrt{2}/a). \tag{2.2-1}$$

The vector joining the extremities of a Brownian walk of N steps (Fig. 2.2.1) is

$$\vec{R}_N = \vec{r}_1 + \vec{r}_2 + \vec{r}_3 + \dots + \vec{r}_N .$$

The *mean* square distance of these walks is linear in N, since, given that each step is of mean length

$$a = \sqrt{\langle \vec{r}_n^2 \rangle} ,$$

and that any two distinct steps are uncorrelated, then

$$R^2 \equiv \langle \vec{R}_N^2 \rangle = \sum_{n,m} \langle \vec{r}_n \cdot \vec{r}_m \rangle = \sum_n \langle \vec{r}_n^2 \rangle = Na^2 . \tag{2.2-2}$$

Therefore, the mean distance traveled varies as the root of the number of steps (see also the diffusion problems in Chap. 5),

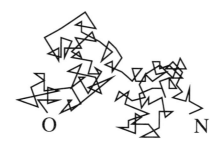

Fig. 2.2.1. Path of a two-dimensional Brownian motion (also called Brownian flight). For a large enough number of steps N, the path may be seen to be dense in the plane: its fractal dimension is D = 2 whatever the dimension of the space in which it is embedded.

$$R = a\sqrt{N} \ .$$

If we take as measure ("mass") the number N of points visited during the Brownian motion or, alternatively, the length $\mathcal{L}$ of the trajectory ON, the "mass"–radius relation

$$N = (R/a)^2 \ \text{or, alternatively,} \ \ \mathcal{L} = Na = R^2/a \qquad (2.2\text{-}3)$$

reveals that the fractal dimension of the Brownian motion is D = 2, and that this is so *whatever the dimension d of the space* in which it is embedded (d ≥ 2).

For d = 1, the Brownian motion cannot have fractal dimension D = 2. In this case the trajectory passes infinitely often within a given distance of an arbitrary point.

The trajectory (path) of a Brownian motion is, of course, much too irregular to represent a coastline (Fig. 2.2.1). Furthermore, a coastline is a curve which does not intersect itself.

Calculation of the probability distribution

Let us now consider the Brownian motion of a particle, and try to determine the probability of finding it at a point $\vec{R}$ at time t, given that it is at the origin at t = 0. Insofar as the time t remains, on average, proportional to the number of steps N, the geometric and dynamical aspects of the situation coincide. (In Chap. 4 we shall tackle problems concerning abnormal waiting times between jumps.)

The *central limit theorem* allows us to obtain the distribution of R_N as N tends to infinity. It is well known that the probability distribution turns out to be Gaussian (a and τ fixed),[4]

$$P(\vec{R},t) \rightarrow \frac{1}{(4\pi\,\mathcal{D}t\,)^{d/2}} \ \exp -\frac{\vec{R}^2}{4\,\mathcal{D}t} \qquad (2.2\text{-}4)$$

with t = Nτ, and the diffusion coefficient, $\mathcal{D} = \langle \vec{r}_n^2 \rangle / 2\,\tau$.

So, the mean square distance is given by

$$R^2 = \langle \vec{R}^2 \rangle = 2\,\mathcal{D}t \ . \qquad (2.2\text{-}5)$$

It can be seen from Eq. (2.2-4) that P is a function of R/√t :

$$P = \frac{1}{R^d} \ f\!\left(\frac{R}{\sqrt{t}}\right) \ .$$

This is a scaling law for the Brownian motion: the probability P dv of finding the particle in a volume dv around $\vec{R}$ is invariant under the simultaneous application of the transformations t → bt and R → λR, with λ = b².

---

[4] A drift $\vec{V}$ due to an exterior field may, in addition, be superimposed on the Brownian motion.

2.2.2 Scalar Brownian motion

Let us now examine another curve, which represents the cumulative gain of someone playing heads or tails against the number of tosses. What is the distribution of the times of return to zero gain for the player?

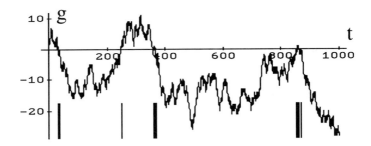

Fig. 2.2.2. *Scalar Brownian motion. The vertical segments at the bottom of the figure indicate the positions on the t- axis of the zeroes of g(t).*

This distribution closely resembles a Levy flight in that it occurs in bursts (the line passes through the axis more frequently when it is close to zero gain than when it is far from it).

Let g be the gain at time t given that it was zero at time t = 0. From above, the probability of having a gain (g > 0) or a loss (g < 0) at the end of a time interval t (the time between two tosses is taken to be unity) is

$$P(g,t) = \frac{1}{\sqrt{2\pi t}} \exp -\frac{g^2}{2t} . \tag{2.2-6}$$

Therefore, the distribution of returning times to the origin (zero gain) follows the law

$$P(0,t) = \frac{1}{\sqrt{2\pi t}} , \tag{2.2-7}$$

so that the *distribution of the times of zero gain is a fractal distribution of dimension D = 1/2.*

To determine the fractal dimension of the times of zero gain, we calculate the number of events occurring on average during an interval of time t, thus

$$N_0 = \int_0^t P(0,t') \, dt' = \int_0^t \frac{dt'}{\sqrt{2\pi t'}} = \sqrt{\frac{2}{\pi}} \, t^{1/2} \propto t^D$$

The distribution of the times of zero gain is a Cantor dust as may be seen from Fig. 2.2.2 (displayed as Cantor bars). Note that the time intervals between tosses need not be equal, a Poisson distribution, say, may equally well be used. The mean time between two jumps must, however, be finite.

A scalar Brownian curve can also be used represent the variation in time of the potential $V_B(t)$ at the terminals of a piece of electrical equipment (presence of Brownian noise).

Finally, note that Brownian motion in d = 3 is the motion of a point each of whose coordinates $\{X_B(t), Y_B(t), Z_B(t)\}$ is a scalar Brownian function.

Returning to mountain reliefs: the *scalar Brownian* curve of Fig. 2.2.2 appears to be a reasonable simulation of a ridge line. We shall now demonstrate that it is possible to generalize this to higher dimensions.

2.2.3 Brownian function of a point

The Brownian function of a point B(P) was defined by Lévy (1948) in the case of a sphere, and then, independently, generalized to the plane by Mandelbrot and Tchensov. The process is as follows: starting with a plane,

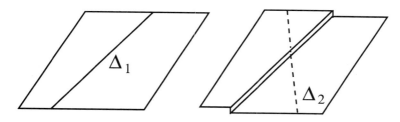

Fig. 2.2.3. Construction of a Brownian function of a point from a plane (this method may be generalized to the sphere and a planetary relief thus constructed).

random steps are created along lines Δ_i of random position and direction (Fig. 2.3.3). By repeating this operation an infinite number of times a fractal surface is generated. This surface is not self-similar (as the vertical direction is privileged), but it is self-affine. We shall return to this concept later in this section.

2.2.4 Fractional Brownian motion

Fractional Brownian motion is one of the most useful mathematical models for describing a number of random fractal structures found in nature. It is an extension of the usual notion of Brownian motion (Mandelbrot and Wallis, 1968). Fractional Brownian motion is represented on a graph by a function which shows the amplitude V_H of the motion as a function of time.

A *fractional Brownian function* $V_H(t)$ is a single valued function of a variable t (possibly representing a time-dependent potential). The variance of its increments $V_H(t_2) - V_H(t_1)$ has a Gaussian distribution:

$$\mathcal{V}(t_2 - t_1) = \Delta V^2(t) = \langle [V_H(t_2) - V_H(t_1)]^2 \rangle = A \, |t_2 - t_1|^{2H}, \qquad (2.2\text{-}8)$$

where $\langle \rangle$ denotes the mean of $V_H(t)$ over many samples, and where H has a value in the range $0 < H < 1$. This function, the mean square increment, is stationary and isotropic (i.e., it depends only on $t_2 - t_1$ and is invariant under t

$\rightarrow$ –t); all the t are statistically equivalent. The particular value H = 1/2 corresponds to ordinary Brownian motion, for which $\Delta V^2 \propto$ t. As with ordinary Brownian motion, V_H (t) is continuous but nowhere differentiable.

We shall see later (in the chapter on transport in fractal media) that the case H $\neq$ 1/2 corresponds to an anomalous diffusion coefficient. Indeed, the diffusion coefficient may be generalized by analogy with expression (2.2-4) (with which it agrees in the case H = 1/2),

$$\mathcal{D}_H = \frac{1}{2} \frac{\partial}{\partial t} \langle V_H(t)^2 \rangle \propto t^{2H-1} . \tag{2.2-9}$$

Thus for H < 1/2, the diffusion becomes increasingly difficult as time passes—there is "subdiffusion", whereas for H > 1/2, this is the reverse—there is "superdiffusion". These anomalous diffusions may be observed in disordered or poorly connected media (H < 1/2), in which the subdiffusion resembles the "ant in a labyrinth" model devised by de Gennes (see Sec. 5.2.2), or in turbulent media (H > 1/2), in which the particle is "superdiffusive" because it is carried by eddies of all sizes, allowing it to travel indefinitely far without changing direction (the trajectory can then be compared to a Levy flight).

Several attempts have been made (especially in relation to the problem of light diffusion by fractals) to make sense of the notion of the "derivative of a fractional Brownian motion" as *fractional Gaussian noise*. The derivative of a normal Brownian motion H = 1/2 corresponds to *uncorrelated white Gaussian noise* so that the Brownian motion is said to have *independent increments* (see Fig. 2.2.4).

Fig. 2.2.4. Graph of white noise, "derivative" (increments) of the first 200 tosses of the Brownian graph of Fig. 2.2.2. The black areas are due to a lack of graphical resolution of sequences of heads (+1) and tails (–1).

Formally, for three times, $t_1 < t_0 < t_2$, $\Delta V_1 = V_H(t_0) - V_H(t_1)$ and $\Delta V_2 = V_H(t_2) - V_H(t_0)$ are statistically independent when H = 1/2. To verify this we calculate

$$\mathcal{V}(2t) = \langle [V_H(t) - V_H(-t)]^2 \rangle = A \mid 2t \mid^{2H} ,$$

and thus $$\mathcal{V}(2t) = 2 \langle V_H(t)^2 \rangle - 2 \langle V_H(t) V_H(-t) \rangle .$$

If we choose $V_H(0) = 0$, then $\langle V_H(t)^2 \rangle = \langle [V_H(t) - V_H(0)]^2 \rangle = A \mid t \mid^{2H}$ and so the *correlation function of the increments* may may be determined:

$$\langle \Delta V_1 \, \Delta V_2 \rangle = \langle \, [V_H(t_0) - V_H(t_1)][V_H(t_2) - V_H(t_0)] \, \rangle = \langle -V_H(-t) \, V_H(t) \, \rangle$$

(taking $t_0 = 0$, $t_1 = -t$, $t_2 = t$).

After normalization, i.e., division by $(\langle \Delta V_1{}^2 \rangle \langle \Delta V_2{}^2 \rangle)^{1/2} = \langle V_H(t)^2 \rangle$, we have

$$\Gamma(t) = \langle -V_H(-t)V_H(t) \rangle \, / \, \langle V_H(t)^2 \rangle = 2^{2H-1} - 1. \qquad (2.2\text{-}10)$$

This function vanishes for $H = 1/2$. When $H > 1/2$, *there is a positive correlation* (the motion is said to be *persistent*) for both the increments of $V_H(t)$ and for its derivative the fractional Brownian noise. When $H < 1/2$, the *increments are negatively correlated* (the motion is said to be *antipersistent*) (Fig. 2.2.5).

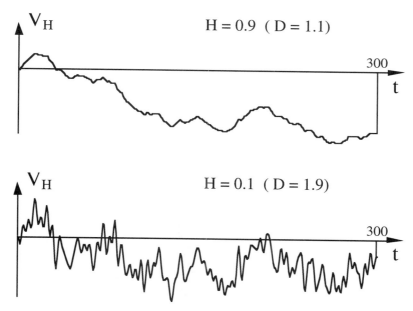

Fig. 2.2.5. Graphs of fractional Brownian functions. In the upper graph the function is persistent (H > 1/2), in the lower one it is antipersistent (H < 1/2); the Brownian case, (H = 1/2) corresponds to Fig. 2.2.2.

This is true at all time scales: like self-similar structures, $V_H(t)$ displays statistical scaling behavior. If t is changed to λt, the increments ΔV_H change by a factor λ^H :

$$\langle \Delta V_H(\lambda t)^2 \rangle \, \propto \, \lambda^{2H} \langle \Delta V_H(t)^2 \rangle,$$

so that, unlike self-similar fractals, the path of $V_H(t)$ requires different scaling factors for the two coordinates (λ for t and λ^H for V_H). t plays a peculiar role since for each value of t there corresponds only one value of V_H.

Such an anisotropic scaling relation is called a *self-affinity* rather than a self-similarity.

Brownian motion constructed from a white noise W(t)

A Brownian motion V_B may be generated by integrating a white noise $W(t)$ (see for instance R. F. Voss, 1989), i.e., a function for which

$$\langle W(t)\, W(t') \rangle = w^2\, \delta(t - t')$$

so that $V_B(t) = \int_{-\infty}^{t} W(t')\, dt'$

or when the motion is discrete, by summing independent jumps or increments,

$$V_B(t) = \sum_{i = -\infty}^{+\infty} A_i\, \Upsilon(t - t_i)$$

where Υ is a step function.

A fractional Brownian motion may also be constructed from a fractional Gaussian noise by introducing a temporal correlation into $W(t)$. We calculate the integral

$$V_H(t) = \frac{1}{\Gamma(H+1/2)} \int_{-\infty}^{t} (t-t')^{H-1/2}\, W(t')\, dt' .$$

These integrals do not exist when $t' \to \infty$, so that, in practice, the increments are calculated between times zero and t (Mandelbrot and van Ness, 1968):

$$\boxed{\Delta V_H(t) = V_H(t) - V_H(0) = \frac{1}{\Gamma(H + 1/2)} \int_{-\infty}^{t} K(t - t')\, W(t')\, dt'} \quad (2.2\text{-}11)$$

where $K(t-t') = (t-t')^{H-1/2}$ if $0 \le t' \le t$,
       $ = (t-t')^{H-1/2} - (-t')^{H-1/2}$ if $t' < 0$.

Noting that $W(\lambda t) = \lambda^{-1/2}\, W(t)$ (derivative of a Brownian motion), it can easily be verified from the previous expression that

$$\Delta V_H(\lambda t) = \lambda^H\, \Delta V_H(t)$$

2.2.5   Self-affine fractals

We have seen in the case of fractional Brownian motion that a structure which is invariant under different scaling laws for different axes is called *self-affine*. In general terms, if for a self-affine fractal curve $V(t)$ in the plane $\{V,t\}$ we consider an interval $\Delta t = 1$ corresponding to a vertical variation $\Delta V = 1$, then V is self-affine if the transformation $\Delta t \to \lambda \Delta t$ transforms $\Delta V \to \lambda^H \Delta V$ with H different from one.

Taking $\lambda = 1/5$ and $\lambda^H = 1/3$ as an example, let us consider the deterministic fractal constructed in the following manner: each diagonal segment of a rectangle is replaced at the following iteration by a broken line made up of five new segments inscribed in a 5x3 rectangle, and whose end points coincide with those of the initial segment (see Fig. 2.2.6).

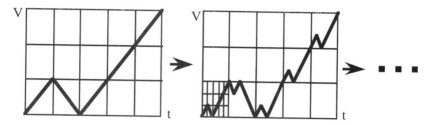

Fig. 2.2.6 Construction of a fractal curve showing an internal affinity. Each segment, the diagonal of a rectangle, is replaced by the generator composed of five new elements.

Relation between D and H for self-affine Brownian motion

It can be seen that this infinitely broken line (initial rectangle as large as required and iterations continuing to infinity) possesses a dilation (or contraction) invariance of factor 5 for the t-axis and factor 3 for the V-axis. Let us calculate its fractal dimension.

If Δt is divided into n parts so that $\Delta t' = 1/n$ (here n = 5, 5^2,...), then ΔV is divided into n^H parts; hence $\Delta V' = 1/n^H$ (here $n^H = 3$, 3^2,...). Using the box-counting method to determine the fractal dimension, we cover the curve portions $\Delta V'$ with $(1/n^H)/(1/n)$ square boxes of side $1/n$ along the V-axis (up to some negligible error of higher order in $1/n$), and repeat the operation n times along the t axis, a total therefore of $N(n) = n \times (n/n^H)$ boxes.

We thus obtain a power law $N(n) = n^{2-H}$ for the number of boxes as a function of their size, which corresponds to (see Sec. 1.2) a *fractal dimension*:

$$\boxed{D = 2 - H} \qquad \text{with } 0 < H < 1 . \qquad (2.2\text{-}12)$$

In the example above, $H = \log 3/\log 5$ hence $D \cong 1.32$. Similarly the self-affine curve of ordinary scalar Brownian motion ($H = 1/2$) has dimension $D = 1.5$ (Fig. 2.2.2). The dimensions of the curves shown in Fig. 2.2.5 are $D = 1.1$ and $D = 1.9$ respectively.

The *zero set* of a fractional Brownian motion is, by definition, the intersection of its graph with the t- axis, i.e., the set of points such that $V_H(t) = 0$ [or more generally such that $V_H(t) - V_H(t_0) = 0$]. In the present case, the zero set is a set of points of topological dimension zero and of fractal dimension $D_0 = D-1 = 1-H$: $0 < D_0 < 1$ so it is a fractal dust.

It is now possible to make the connection between the Brownian motion of a player's gain whose dimension is $D = 3/2$, and the distribution of the times of zero gain which is a Cantor dust of dimension $D_0 = D-1 = 1/2$ (Fig. 2.2.2).

Generalization to d dimensions

A fractal Brownian motion in a d-dimensional Euclidean space may similarly be defined such that

$$\langle[V_H(x'_1\ldots x'_d) - V_H(x_1\ldots x_d)]^2\rangle \propto [\,|\,x'_1-x_1\,|^2 + \ldots\,|\,x'_d-x_d\,|^2\,]^H\,. \quad (2.2\text{-}13)$$

The hypersurface generated by the points $\{x_1,\ldots x_d, V_H\}$ is a self-affine fractal of fractal dimension:

$$\boxed{D = d + 1 - H} \qquad\qquad (2.2\text{-}14)$$

and of topological dimension $d_T = d$.

Moreover, whereas $V_H(x_1, x_2, \ldots, x_d)$ is self-affine, it can easily be shown that the *zero set*, that is, the intersection with a (hyper-)plane parallel to the $\{x_1,\ldots x_d\}$ plane, is *self-similar*. Its dimension is,

$$\boxed{D_0 = d - H}\,. \qquad\qquad (2.2\text{-}15)$$

Remark: If the dividers' method is used to calculate the dimension, a completely different value is found, known as the *latent dimension*. To determine this dimension, D_c, corresponding to steps ε, we write

$$\varepsilon^2 = \lambda^2 \Delta t^2 + \lambda^{2H} \Delta V^2\,.$$

When $\Delta V / \Delta t$ is sufficiently large, $\varepsilon \propto \lambda^H$. The total length of the measure with step length ε is $N\varepsilon$, where N is the number of steps. For a graph of size T where $t \in [0,T]$, we have $N = T/(\lambda \Delta t)$, and $L = N \propto \lambda^{H-1} \propto \varepsilon^{1-1/H}$.

Referring to Sec. 1.4, Eq. (1.4-1) we see that the latent fractal dimension is

$$D_c = 1 / H.$$

Global dimension

The "local" dimension $D = d-H$, which we have just determined, is not the only dimension associated with self-affine fractals. A mountainous massif may be taken as a good representation of a self-affine fractal in a space of $d(=2)+1$ dimensions. From what we have seen, the fractal dimension observed locally (for a hiker contemplating the scene) is given by $D = 3-H$. However, flown over at a high enough altitude, only the global aspect of the ground surface is apparent and the *global dimension* will be $D = 2$ (Mandelbrot, in Pietronero and Tossati, 1986).

Let us try to make this idea more precise. We know that fluctuations in $V(x_1,\ldots,x_d)$ vary with distance R as R^H (by the definition of fractional Brownian motion). So, if we measure the fractal dimension with square boxes of size $\varepsilon > \chi$, where χ is such that $\chi \cong \chi^H$, the effect of varying the altitude will no longer be significant, since, H being less than 1, the amplitude ε^H of the fluctuations of the self-affine curve grows more slowly than ε. In this case the dimension obtained will simply be $D = d$. In the example shown in Fig. 2.2.7 (for which $d = 1$), the fluctuations in t^H are represented by the enveloping curve (dashed curve in Fig. 2.2.7). The use of boxes of size equal to (or greater than) χ leads to a dimension equal to one. This is because the number of boxes covering the object is then proportional to t. Taking for example an interval $\Delta t = 4\chi$, the section of the curve $V_H(t)$ is covered with 4 squares of side $\varepsilon = \chi$ and with 2 squares of side $\varepsilon = 2\chi$, that is, approximately 4/n for $\varepsilon = n\chi$, whereas 12 squares are needed if $\varepsilon = \chi/2$, and of the order of $4n^{1.32}$ would be needed for $\varepsilon = \chi/n$, for large n.

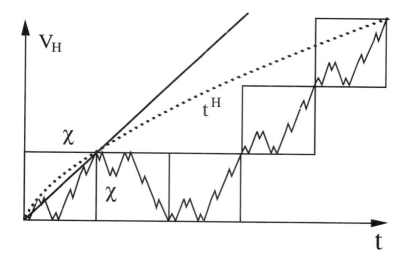

Fig. 2.2.7. *Local dimension and global dimension of a self-affine structure. At distances smaller than χ, the structure is fractal of dimension 1.32..., but at greater distances the structure becomes one-dimensional since $0< H<1$.*

Dimension of the path

We saw in Sec. 2.2.1 that the path of a Brownian motion (or Brownian flight) was, whatever the dimension d of the space, a *self-similar* structure of dimension $D = 2$. The problem now arises of determining the fractal dimension of the path of a fractional Brownian motion. By definition, such a path is composed of t steps of random direction and of length $\Delta V_H(t)$.

By using the box-counting method (but the other methods would give the same dimension), the dimension of the path can easily be seen to be exactly equal to the latent dimension obtained for the Brownian function by the dividers' method (cf. the remark above)

$$\boxed{D(\text{path}) = 1/H}. \tag{2.2-16a}$$

In particular, this formula can be used to confirm that Brownian motion, which corresponds to $H = 1/2$, traces out a path of dimension 2, and that very persistent Brownian motion ($H \cong 1$) has dimension $D \cong 1$.

It should also be noted that multiple points become dominant when $H < 1/d$, the path then becoming dense in the space

$$\boxed{D(\text{path}) = d \ \text{if} \ H < 1/d} . \tag{2.2-16b}$$

Other self-affine sets

The previous discussion was restricted to the description of self-affine curves relating to fractional Brownian motion, but internal affinity pertains equally to objects possessing very general structures. The generator defined by Fig. 2.2.8 gives an example of a self-affine structure which is not a curve $V(t)$. The fractal dimension, calculated by the box-counting method in a similar way to the fractal in Fig. 2.2.6, has, in general, a value different from the Hausdorff dimension.

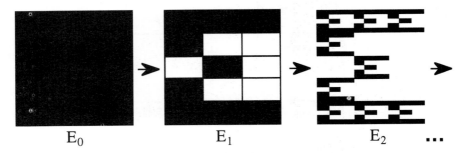

E_0 $\qquad\qquad\qquad$ E_1 $\qquad\qquad\qquad$ E_2 $\quad$ ...

Fig. 2.2.8. *First two iterations of a self-affine fractal: E_0 is the initiator, E_1 is the generator (identical to the first iteration), E_2 corresponds to the second iteration, etc.*

In general terms, if we take a unit square E_0 (initiator) and divide it into a network of $p \times q$ rectangles with sides $1/p$ and $1/q$ ($p<q$), and if we select from this network a set E_1 (generator) of rectangles, the number of rectangles selected from each column $1 \leq j \leq p$ denoted by N_j, then after an infinite number of iterations resulting in the limit set E, we have (Falconer, 1990) :
— for the Hausdorff dimension,

$$\dim E = \log\left(\sum_{j=1}^{p} N_j^{\log p/\log q}\right)\frac{1}{\log p} \ ;$$

— for the box-counting dimension ,

$$\Delta(E) = \frac{\log p_1}{\log p} + \log\left(\frac{1}{p_1}\sum_{j=1}^{p} N_j\right)\frac{1}{\log q} \ ,$$

where p_1 is the number of columns containing at least one rectangle of E_1. In the figure above, $p = 3$, $q = 5$ and $\dim E = 1.675...$ and $\Delta(E) = 1.683...$

For a more detailed discussion and various examples of self-affine fractals refer to Falconer (1990), Barnsley (1988), and Mandelbrot (1982).

2.2.6 Mountainous reliefs

Without entering into a physical explanation of the fractal nature of a

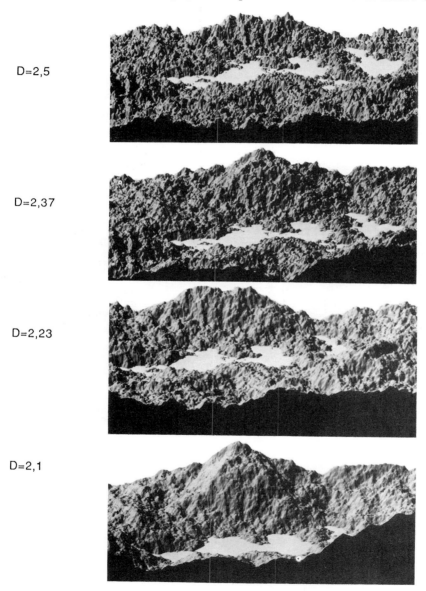

D=2,5

D=2,37

D=2,23

D=2,1

Fig. 2.2.9. Example of fractal construction using affine fractional Brownian functions: Constructions simulating reliefs in 3D space (Mandelbrot, 1982).

mountainous relief, which is in any case not yet clearly understood (see Sec. 2.2.9 on fractures), we shall limit ourselves here to giving a visual impression. The four images in Fig. 2.2.9 simulate a mountain relief generated by a Brownian function of a point with a given fractal dimension ranging from 2.1 to 2.5. They can be constructed by Fourier transforming a $1/f^\beta$ distribution.

Artificial clouds can also be created using this construction in $d = 4$ (three spatial coordinates and a parameter representing the water vapor density), the "zero set" then being a section in $d = 3$ which simulates cloud formations very accurately (see Sec. 2.2.8). It turns out that many natural phenomena are observed to have an H close to 0.8.

A spectacular sequence of magnifications of a fractal landscape is shown in Fig 2.2.10. Such a sequence is even more impressive in an animated presentation.

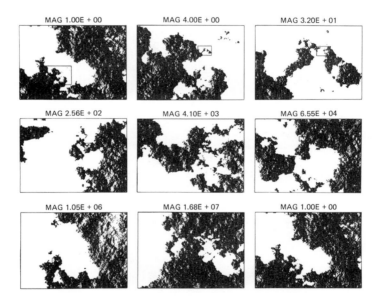

Fig. 2.2.10. This sequence represents a statistically self-similar, fractal landscape (the landscape itself has dimension D = 2.2) for several enlargements ranging from a factor 1 to a factor greater than 16 millions. Note that the statistical generation has been carefully programmed to recapture the initial image (Voss, in Peitgen and Saupe, 1988)

2.2.7   Spectral density of fractional Brownian motion, the spectral exponent β

Let $V(t)$ be a random function. It can be characterized by its *spectral density* $S_v(f)$. $S_v(f)$ is the mean square of the Fourier transform of V, $V(f)$, per unit width of band. For example, at the exit of a $V(f)$ filter of width Δf around f (see, e.g., A. van der Ziel, 1970):

$$S_v(f) = \langle |V(f)|^2 \rangle \, / \, \Delta f \quad . \tag{2.2-17a}$$

S_v therefore provides information about the correlation time characterizing $V(t)$. If the Fourier transform $V(f,T)$ of $V(t)$ for $0 < t < T$ is defined by

$$V(f,T) = \frac{1}{T} \int_0^T V(t) \, e^{2\pi i f t} \, dt$$

then $$S_v(f) \underset{T \to \infty}{\propto} T \langle |V(f,T)|^2 \rangle \quad . \tag{2.2-17b}$$

The spectral density can also be related to the *correlation function* at two points of $V(t)$:

$$G_v(\tau) = \langle V(t) \, V(t+\tau) \rangle - \langle V(t) \rangle^2 \quad . \tag{2.2-18}$$

When the Wiener–Khintchine theorem applies (i.e., when the correlation function converges to zero sufficiently rapidly at infinity) we have

$$G_v(\tau) = \int_0^\infty S_v(f) \, \cos(2\pi f \tau) \, df \quad . \tag{2.2-19}$$

In particular for Gaussian noise, $S_v(f) = $ constant and $G_v(\tau) = \Delta V^2 \, \delta(\tau)$ is completely uncorrelated. Similarly for a power law,

we have
$$S_v(f) \propto 1/f^\beta \quad \text{with } 0 < \beta < 1$$
$$G_v(\tau) \propto \tau^{\beta - 1}.$$

Furthermore, from the equation

$$\langle |V(t+\tau) - V(t)|^2 \rangle = 2 \, [\, \langle V^2 \rangle - \langle V \rangle^2] - 2 G_v(\tau) \quad ,$$

the relationship

$$\beta = 2H + 1 \quad . \tag{2.2-20}$$

may be deduced for fractional Brownian motion, $V_H(t)$.

Brownian motion $V_H(t)$ in d dimensions has fractal dimension D and spectral density $S_v(f) \propto 1/f^\beta$ such that $D = d + 1 - H = d + (3 - \beta)/2$.

As can be seen from the previous discussion, Fourier transforms may be used advantageously to construct fractional Brownian motions.

Fractal music

Here we mention an interesting discovery of Voss and Clarke (Voss, 1988) concerning the spectral density of music composed in different civilizations and in different eras: remarkably it is approximately $1/f$ (Fig. 2.2.11).

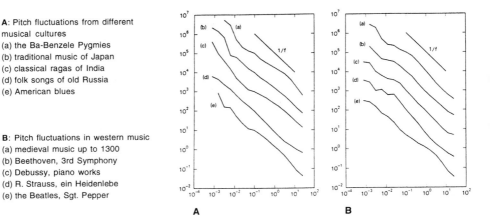

A: Pitch fluctuations from different musical cultures
(a) the Ba-Benzele Pygmies
(b) traditional music of Japan
(c) classical ragas of India
(d) folk songs of old Russia
(e) American blues

B: Pitch fluctuations in western music
(a) medieval music up to 1300
(b) Beethoven, 3rd Symphony
(c) Debussy, piano works
(d) R. Strauss, ein Heidenlebe
(e) the Beatles, Sgt. Pepper

Fig. 2.2.11. Frequency spectrum of a variety of music from different eras and varied cultures (f in Hertz). They are fairly close to 1/f behavior (after Voss, 1988).

Random music can be created without much difficulty: white noise[5] sounds too random, Brownian $(1/f^2)$ noise too correlated and it seems a little monotonous, but 1/f noise is more pleasant to listen to (Fig. 2.2.12). Apparently, the 1/f spectral distribution suits our natural physiology. Of course this is only one parameter; to create music there are thousands of others to take into account!

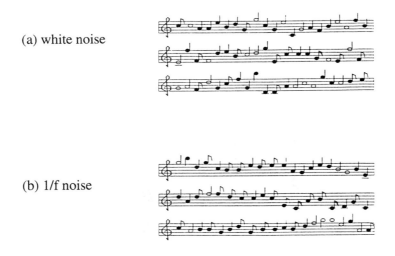

(a) white noise

(b) 1/f noise

---

[5] White noise over a large enough spectrum of frequencies resembles the noise of steam escaping from a steam engine under pressure. For random music the frequency band is much narrower as the usual scale is used.

(c) Brownian

Fig. 2.2.12. Three examples of random computer generated music (Voss, 1988).

2.2.8   Clouds

In 1982 Lovejoy studied the shape of clouds and showed that they have a fractal structure over more than four orders of magnitude (from 3 to 3 x 10^4 km). The relationship which gives the area as a function of the perimeter is a power law whose exponent is close to 1.33 (Fig. 2.2.13). The measurements were made from radar and satellite images. In addition, analysis of radar data (Rys and Waldvogel, 1986) on (very convective) hail-storm clouds gives a Hausdorff dimension for their perimeter of about $D = 1.36 \pm 0.1$. The fractal nature of clouds allows a very convincing and useful numerical computer simulation to be made of them. The underlying physical causes of this nature are related to turbulent diffusion (see Sec. 2.3).

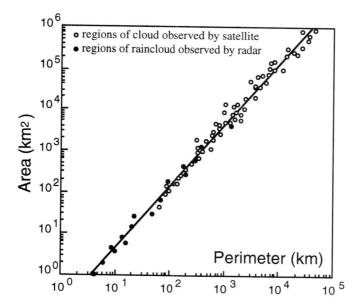

Fig. 2.2.13. Analysis of the relationship between the area of clouds (varying as λ^2 where λ is a characteristic length ranging from the order of 1 km to several hundred km) and their perimeter. (Lovejoy, 1982).

2.2.9 Fractures

It is easy to imagine that if no identifiable object is there to fix the scale, an enlargement of the surface of a broken stone may be indiscernible from a photograph of a cliff or mountain face. Dilational invariance is clearly present here. Semiempirical arguments and numerous measurements suggest that fracture surfaces in metal may have a fractal nature. These characteristics may be compared with the fractal structures of mountain reliefs and also share features with dielectric breakdown as well as with diffusion-limited aggregation (which we shall discuss in Sec. 3.4.5 and Sec. 4.2).

Fractures may be grouped, roughly speaking, into two classes: fragile ruptures in which the material breaks without deforming, like a piece of crockery falling on a hard floor; and ductile ruptures in which the material deforms significantly before breaking.

Experimental simulations of fractures of thin layers have been performed by Skjeltorp in Norway (Skjeltorp, 1988). These studies are very interesting because of the way they have been carried out: the experiment consists in making single layers of microspheres (spheres of sulphurized[6] polystyrene of diameter 3.4 ± 0.03 µm) dispersed in water. By confining them between two plane glass plates, it is possible to form a two-dimensional polycrystal for which the size of the monocrystalline "grains" is reasonably large (10^5 to 10^6 balls). If a layer like this is dried very slowly, the diameter of the spheres contracts to 2.7 µm. Forces between spheres emerge as they have a tendency to remain stuck to each other, whereas the forces between the spheres and plates of glass remain negligible.

As no prefracture has been made at the edge of the layer, cracks arising from flaws appear over the whole sample. Their number increases during the drying process by a succession of hierarchical branchings. The stresses in the system diminish within regions not yet fractured, and because of this so do the widths of the cracks there. Figures 2.2.14 (a–d) show various enlargements of the cracked layer. The distribution of the cracks is fractal over more than two orders of magnitude (Fig. 2.2.15) as analysis by the box-counting method shows (N boxes of size ε):

$$N \propto \varepsilon^{-D}$$

Fractures are very often initiated by a surface flaw (prefracture, etc.); an isolated fracture then crosses the sample. In the case of layers, the fracture is a curve whose structure is not revealed by the previous analysis (which concerns a fracturing of a body). In the case of a volume, this type of fracture is a surface similar to a self-affine fractal. Quantitive analysis in terms of fractals has only started in the last few years. Mandelbrot, Passoja, and Paullay (1984) were the first to use the concept of fractals to describe fractured metallic surfaces. Other measurements on surfaces of titanium were then carried out by

---

[6] The spheres are sulfurized to make them contract during the drying process.

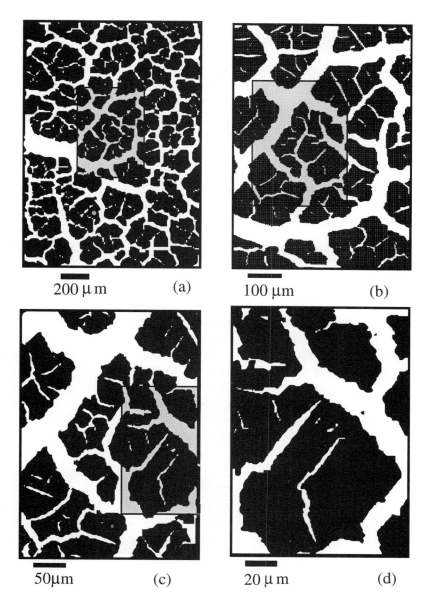

Fig. 2.2.14. Final structure of a fractured monolayer at various magnifications (Skjeltorp, 1988).

Pande et al. (1987). The roughness of fractured surfaces has also been used by Davidson (1989) as a measure of the hardness and the resistance to fatigue of materials. The most recent experiments have been carried out on samples of

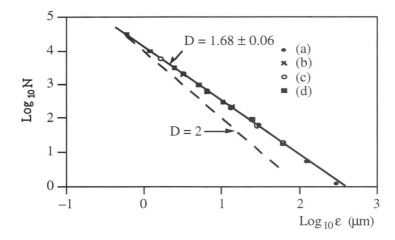

Fig. 2.2.15. Calculation of the fractal dimension of the fractured surface of a monolayer of microspheres; the box-counting method has been used over the four magnifications (a–d) of Fig. 2.2.14 (after Skjeltorp, 1988).

metals with different hardnesses (aluminium alloys with different heat treatments). Curiously, in these experiments the same fractal dimension $D \cong 2.2$ was obtained for the different samples, and this has led to pose the problem of finding the relationship between hardness and fractal dimension as suggested by Davidson. Using the same methods as Mandelbrot et al. (1984), the fractal dimension is measured by making a series of cuts parallel to the fracture after it has been coated with nickel, thereby enabling it to be polished without altering its surface (E. Bouchaud et al., 1990) (Fig. 2.2.16). In each case the experimental conditions (anisotropic and prefracturation constraints

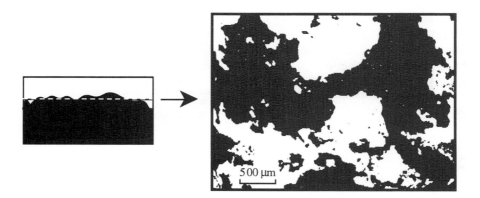

Fig. 2.2.16 Section of a fracture in an aluminium alloy (in black). The fracture is covered in nickel (in white) before the section is made (Bouchaud et al., 1990).

applied) cause the fractured part to display self-affine symmetry. The fractal dimensions in the planes parallel to the fracture have been measured both by the box-counting method and by the *perimeter-area* relation. This relationship relies on the fact that a section of the surface S of the fractured material (in black in Fig. 2.2.16) is not fractal and so varies as R^2, where R is the mean diameter of S. The perimeter P of this section varies as R^{D-1}, where D is the dimension of the fracture, and hence we have a relationship between the perimeter and the area, $P \propto S^{(D-1)/2}$, which allows us to calculate D (see also Fig. 2.2.13 where this method is applied).

Unfortunately, at the present time all these studies are essentially descriptive: the physical reasons for the fractality of ruptures are not at all well understood, especially in three dimensions, and no clear relationship between the mechanical properties and the fractal dimension is known. The principal difficulty arises from the fact that, even for an ideal homogeneous material, a complicated field of nonlocal forces develops as the material starts to break.

The starting point for a theoretical model is the equation for the strains $\vec{u}$ in a continuous elastic medium,

$$(\lambda+\mu) \, \vec{\nabla} \, (\vec{\nabla}\cdot\vec{u}) + \mu \, \Delta\vec{u} = 0, \qquad (2.2\text{-}21)$$

λ and μ being the Lamé coefficients. The assumption is then made that the probability of a given part of the sample rupturing is proportional to a power η of the force to which it is submitted (Louis *et al.*, 1986).

The scaling laws of fractures have been obtained numerically, but only in two-dimensional systems (de Arcangélis *et al.*, 1989). The model is composed of a network of fragile bonds, each of which will break when the force exerted on it exceeds a certain threshold. The breaking thresholds of the bonds are chosen randomly following a given probability distribution. The remarkable thing these authors have found is that this fragile material has a fractal dimension which is independent of the distribution rule for the breaking thresholds, that is to say, independent of the nature of the (fragile) material, in agreement with the experimental results of Bouchaud *et al.* concerning ductile materials. Furthermore, Guinea and Louis have discovered that the broken bonds of an elastic network form a fractal of dimension D = 1.25. However, it is too early at the moment to assert that fractures obey universal laws.

At the time of writing, no conclusive study has been carried out in three dimensions.

2.3 Turbulence and chaos

The majority of flows appearing in nature are turbulent. This is true of the upper layer of the earth's atmosphere, the "*jet stream*s" of the upper troposphere, cumulus clouds, and ocean currents such as the Gulf Stream. It is also true of the photosphere of the sun and other similar stars, interstellar gaseous nebulae, and, at a smaller scale, of boundary layers on an aeroplane's wings, river currents, and rising smoke. Turbulence is used to mix or homogenize fluids and to accelerate chemical reactions. In the last few years, various manifestations of turbulence have been the object of numerous studies. We shall make only a brief attempt here to try to understand its fractal nature. Systems possessing a large number of degrees of freedom give rise to "strong turbulence", but one of the great discoveries of the last decades has been the realisation that only three degrees of freedom are required to let chaos appear.[8] (Ruelle and Takens, 1971). "Weak turbulence" is generated in this way. Simple, but nonlinear, equations are at the root of this type of *chaos*, which is then said to be *deterministic* (Schuster, 1984; Bergé et al., 1988), since, at least in principle, the state of the system at any moment may be calculated given the initial state and the equations. Later we shall examine in a little more detail examples of phenomena generating deterministic chaos.

2.3.1   Fractal models of developed turbulence

In a low-viscosity fluid, the motion of a small portion of this fluid is subject to such little resistance that a minute perturbation may be amplified and, by a series of instabilities, may lead to disordered motion, qualified as turbulent.

Turbulence arises from the interaction of a large number of eddies together with a rapid increase in the fluid's *vorticity* (characterizing the eddies) whose intensity is denoted $|\vec{\omega}|$,

$$\vec{\omega} = \overrightarrow{\text{curl}} \ \vec{v} \qquad (2.3\text{-}1)$$

$\vec{v}$ being the velocity field in the fluid. This velocity field is itself governed by the Navier–Stokes equation (which is simply Newton's equation, $\vec{F} = m \, d\vec{v}/dt$, applied to an element of fluid). When the eddies extend to all scales, it is said to be a case of *developed turbulence.* This idea of a hierarchy of eddies was portrayed by Richardson in 1922 in the words of a poem inspired by one of Jonathan Swift's poems (cited in the foreword):

> Big whorls have little whorls,
> Which feed on their velocity;
> And little whorls have lesser whorls,
> And so on to viscosity.

The mechanism presumed to be at the root of this hierarchical structuring (but this is still only a conjecture) is based on the fact that eddies of a given size destabilize thereby producing eddies of smaller sizes. This continues until a minimum size when the eddies become stable. The corresponding scale at which the viscosity is sufficient to dissipate the energy is called the

---

[8] For physicists working in this area, the word "chaos" has a fairly precise meaning: it concerns the temporally unpredictable aspect of phenomena (then called chaotic). The word "turbulence" is used rather to describe spatio-temporal behavior.

Kolmogorov scale. The parameter controlling the turbulence is the Reynolds number Re,

$$\text{Re} = UL/v_0 , \tag{2.3-2}$$

where U is a characteristic speed of the fluid, L the size of the object generating the turbulence, and v_0 the kinematic viscosity of the fluid ($v_0 = \mu/\rho$, where μ is the viscosity and ρ the density). The Navier-Stokes equation may then be written

$$\rho \frac{\partial \vec{v}}{\partial t} + \rho (\vec{v} \cdot \overrightarrow{\text{grad}})\vec{v} + \vec{\nabla} p = \vec{f} + \mu \Delta \vec{v} ,$$

$$\text{div } \vec{v} = 0. \tag{2.3-3}$$

Initial and boundary conditions must be added to these equations.

The second equation expresses the conservation of matter. It is assumed that all speeds are small relative to the speed of sound; $\vec{f}$ is the external force applied, p is the pressure in the fluid, and ρ its density. In practice, the

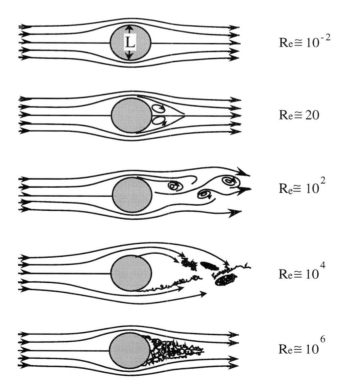

Fig. 2.3.1. Passage to turbulence behind a cylinder of size L, as the velocity U (thus also Re) is progressively increased [after Paladin and Vulpiani (1987)].

Navier–Stokes equation depends only on R_e, via the relationship between the advection term $(\vec{v}\cdot\overrightarrow{\mathrm{grad}})\vec{v}$, and the dissipation term $(\mu/\rho)\Delta\vec{v}$. When the viscosity vanishes ($R_e = \infty$), the equation becomes an Euler equation, for which energy is conserved. The Navier–Stokes equation will not be analyzed any further here, but we should mention that an equation of motion for the eddies may be found by combining[9] it with the equation defining $\vec{\omega}$. It may then be observed that the behavior of *two-dimensional turbulence* is very special since in two dimensions the vector $\vec{\omega}$ remains orthogonal to the velocity field. This point is important in, for example, questions concerning atmospheric movement: turbulence is, roughly speaking, three-dimensional up to a scale of 100 km, then two-dimensional from 100 to 1000 km, while beyond 1000 km the sphericity of the earth comes into play. The large vortical structures of the earth's atmosphere have sizes of several thousands of kilometers, while the scale at which dissipation occurs (Kolmogorov scale) is in the order of millimeters.

Turbulence becomes fully developed for values of the Reynolds' number above a critical value $R_{e_{crit}}$ ($\cong 2000$ in Reynolds' experiments[10]), turbulence itself occurring for values of R_e between roughly 10 and 1000, according to the geometry, and occurring after the laminar phase which is present at low values of R_e. In the laminar phase small fluctuations (residual turbulence) always exist, but they are rapidly damped out. In the turbulent phase, on the other hand, these fluctuations are amplified and then inextricably interact with each other. This happens when the damping time (which is in L^2/v_0) of a perturbation becomes longer than the time needed for a particle to travel the length of the object, of size L, producing the perturbation (i.e., L/U). Behind a grill, developed turbulence is obtained for lower values of $R_{e_{crit}}$ than, say, behind a cylinder. Moreover, in this case the turbulence is relatively homogeneous and isotropic.

The Kolmogorov model (1941)

In 1941, the Russian mathematician Kolmogorov proposed a model of three-dimensional developed turbulence in which the eddies belong to a structure that is arranged hierarchically according to size. The energy is injected into the eddies of largest size then transferred from eddy to eddy in a cascade of decreasing sizes down to the smallest size where the energy is dissipated (the Kolmogorov scale). The theory of Kolmogorov-Obukhov (1941) predicts in particular an energy spectrum in $k^{-5/3}$, where k is the wave vector of the Fourier modes of the velocity field (the exponent 5/3 is easily obtained by simple dimension considerations in the equations). These models are not dissimilar to the cascades of Fournier and Hoyle (Sec. 2.1.2). We start from

---

[9] The pressure term is eliminated when taking the curl of the equation

[10] The critical value depends on the experiment and, in particular, on the level of residual turbulence. By reducing this level, values of $R_{e_{crit}}$ of the order of 10^5 may be arrived at.

the hypothesis that an eddy gives rise to N sub-eddies, r times smaller in size, at the heart of which all the dissipation is concentrated. This process is then repeated suggesting a fractal dimension D = log N /log r. For turbulence to develop it is necessary that D > 2, so that a cut made at random through the space has a nonzero probability of intersecting the support of the turbulence (Mandelbrot, 1974, 1975b). Mandelbrot also suggested (in 1976) that the singularities of the Navier–Stokes and Euler equations, which govern fluid dynamics, may have a fractal structure (this would explain the nature of turbulence), but at the present time it is still not known whether these equations develop singularities after a finite time.

Frisch, Sulem, and Nelkin (1979) put forward the idea that the dissipation is concentrated in a domain of nonintegral fractal dimension and proposed the *β model* in which the energy flux[11] is transferred to a fixed fraction β of the eddies of smaller size (see the construction[12] in Sec. 1.4.2). However, the experimental data relating to the moments of the velocity fluctuations seems to show that the scaling laws at small distances cannot be described by a homogeneous fractal (i.e., constructed by rules relating the statistical properties at a certain scale to those at a much larger scale). So Benzi *et al.* (1984) introduced, on the basis of the work of Novikov, Stewart, and Kraichnan, heterogeneous fractals, of the type described earlier in Sec. 1.4.2, in which the rules relating two levels in the cascade are established according to a given probability law *(random β model)*.

The fractal dimension of the support of the dissipation is then of the form

$$D = d + \log \langle ß \rangle / \log 2,$$

$\langle ß \rangle$ denoting the mean of the branching probabilities β (Sec.1.4.2). A more detailed discussion of this model may be found in an article written by G. Paladin and A. Vulpiani (1987) (see also Vulpiani in Pietronero, 1989). We should mention, however, the important fact that a heterogeneous fractal structure has a *multifractal* nature which, as we shall see later on, allows a more precise comparison between theory and experiment to be made.

Difficulties with theoretical studies stem from the fact that turbulent flow is an open thermodynamic system, i.e., not isolated from the exterior due to the presence of forces acting on the flow: at a large scale, external forces, and, at a small scale, viscosity. Only at intermediate scales may it be supposed that the energy is transferred conservatively between the different degrees of freedom. Among examples of turbulent flow, turbulence behind a grill lends itself fairly well to theoretical idealizations (homogeneous and isotropic

---

[11] All these models concern three-dimensional turbulence. In two dimensions due to the orthogonality between vorticity and velocity fields, the conserved quantity is the mean square of the vorticity (or enstrophy), so that we then have a cascade of enstrophy giving in this case a behavior in k^{-3}.

[12] The example in Sec. 1.4.2 corresponds to $N = r^d$ β and r = 2. The β model is fairly similar to the *"absolute curdling"* model proposed by Mandelbrot (Mandelbrot, 1976).

turbulence, β models) because here it may be assumed that there is no extraction of energy by the turbulence from the mean flow, and that the grill homogenizes the interactions between eddies. Turbulence of coherent structures (atmospheric movements are one example, see also Fig. 2.3.4) is trickier to analyze.

In a similar vein, a model of simple multifractal cascade has been proposed by Meneveau and Sreenivasan (1987a). We shall examine this very simple one-dimensional model as it is can be used to describe reasonably accurately the multifractal behavior of the rate of dissipation ε.

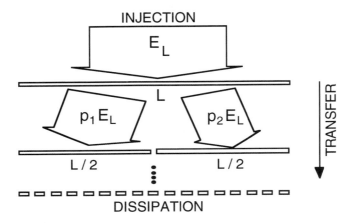

Fig. 2.3.2. One-dimensional version of Meneveau and Sreenivasan's model of cascading eddies (1987). The kinetic energy flux E_L injected into the system is divided into random fractions $\{p_1, p_2\}$ at each level of the hierarchy down to the Kolmogorov scale η.

The model is based on the supposition that the energy injected into a one-dimensional section L of a system (in d = 3) divides into two subregions of size L/2 according to the probabilities p_1 and p_2, and so on, until a minimum size η (Kolmogorov scale) is reached where the energy dissipates (Fig. 2.3.2). This type of model, a *binomial fractal measure*, appeared earlier. Meneveau and Sreenivasan have used it to interpret measures over one-dimensional sections of various flows of developed turbulence (turbulence produced by a grill, slipstream behind a circular cylinder, boundary layers, atmospheric turbulence). The results are shown in Fig. 2.3.3. To be precise, the generalized dimension D_q, which suffices to describe the multifractal structure completely, has been plotted (it can be seen that $D_0 = 1$, as the support of the measure is a line). It is in good agreement with the binomial model in which the value of p_1 is set at 0.7. The horizontal dot-dash line corresponds to the β model (not random, therefore not multifractal, so all the D_q are equal to D < 1)

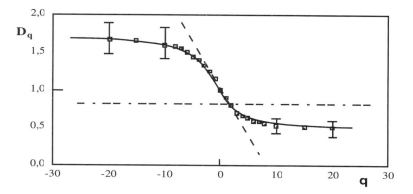

Fig. 2.3.3. *Generalized dimensions of a one-dimensional section of the dissipating field in turbulent flows [Meneveau and Sreenivasan (1987a)]. The manner in which these experiments were carried out is explained in Meneveau and Sreenivasan (1987b).*

and the dashed line to the log-normal[13] model. However, this binomial model is still too simple and may lead to incorrect interpretations. The problem was taken up again in 1988 by the same authors in collaboration with Prasad. The multifractal spectrum f(α), including its latent part which involves negative fractal dimensions, has been found in several types of experimental situation.

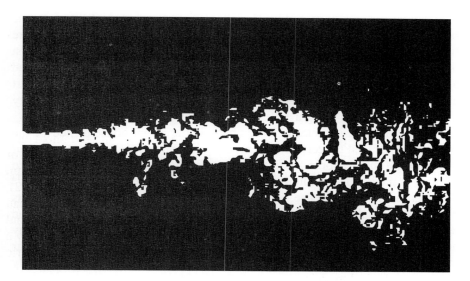

Fig. 2.3.4. *Plane section of an axisymmetric jet obtained by exciting the fluorescence of the injected fluid [after Dimotakis et al. (1981)]. The white regions indicate the distribution of the injected material, not that of the velocity.*

---

[13] A random variable V is said to follow a log-normal distribution if its statistics are completely determined by the two moments $\langle \log V \rangle$ and $\langle (\log V - \langle \log V \rangle)^2 \rangle$.

These results along with a review of some earlier approaches have been published in a recent, extensive article (Méneveau and Sreenivasan, 1991).

> To determine experimentally the fractal dimension, or more generally the multifractal structure, of a three-dimensional turbulent region, it is helpful to separate the regions of a plane where vorticity is present from those where the flow is purely laminar. In Fig. 2.3.4 the fluid is illuminated by a plane laser beam, which shows the distribution of turbulent fluid in a plane cross-section.

> We should mention here a very pretty geometric study carried out by H. Herrmann et al. (1990), who have studied the possibility of disks rolling against each other without friction in such a way that all the space within a band or a circle is filled. They showed that this was possible with a fractal distribution of disks. Their initial aim was to study the absence of energy dissipation in certain types of relative motion of tectonic plates, but the study may also be applied to turbulence where dissipation only occurs at the smallest scales. The model leads to laws in k–5/3 similar to that of Kolmogorov.

2.3.2 Deterministic chaos in dissipative systems

Among the fundamental discoveries of the last 20 years, the appearance of chaos in very simple (but, of course, nonlinear) systems holds an important place. The study of *deterministic chaos* has been considerably extended in recent years: although Poincaré had already noticed in 1892 that certain mechanical systems governed by Hamilton's equations could display chaotic behavior, it was only (in 1963) when the meteorologist Lorenz became aware that a system controlled by laws as simple as a system of three coupled first order differential equations without the interference of external noise could lead to completely chaotic trajectories. Lorenz had discovered one of the first examples of deterministic chaos in dissipative systems (although the discovery took some time to be recognized).

The existence of deterministic chaos is the direct consequence of the property possessed by certain systems of being *extremely sensitive to initial conditions*: a small error in the initial values that have been chosen gives rise to an error in predicting the future evolution that grows exponentially with time.[14] There is then, in some sense, a loss of memory, since after a certain time has elapsed the initial conditions cannot be recovered. When this sort of evolution takes place, whatever the initial conditions, it is said to be chaotic. Many nonlinear systems behave chaotically. To mention just a few: the nonlinear forced pendulum, lasers, various chemical reactions, three or more body systems, biological models of population dynamics, etc.

Among these problems, frictionless systems, known as *conservative* or *Hamiltonian*, must be clearly distinguished from so-called *dissipative* systems in which there is internal friction. Only the latter possess an *attractor* due to the

---

[14] Lorenz gave the example of a butterfly beating its wings in Australia which by changing the development of the winds was the cause of a cyclone in the Antilles.

fact that dissipation causes there to be a stationary limit in the long term, when energy is injected into them from outside (forced systems). These systems have been studied (see, e.g., Smale 1980; Gleick, 1987) by Smale, who is the inventor of the so-called "horseshoe" model of strange attractors. On the other hand, Hamiltonian systems, because of the absence of dissipation, retain the memory of their initial conditions, to which they may be extremely sensitive. They must therefore be treated accordingly.

Dissipative systems

Of these systems, chaos only appears in those that are open, that is, those experiencing an external force which allows the energy lost through dissipation to be reinjected. Indeed, for a closed system dissipation leads to a state of equilibrium, and so it becomes stable. If we consider only dissipative systems, at least three routes towards chaos have been discovered :

(i) by successive *bifurcations* (Feigenbaum, Coullet, and Tresser);

(ii) by passing from periodic attractors to *strange attractors* (by a finite number of bifurcations according to the scenario of Ruelle, Takens, and Newhouse). Here there is a transition from a quasiperiodic situation to chaos;

(iii) by the appearance of an intermittent dynamical situation (*intermittency* as described by Pomeau and Manneville).

As with other topics that extend beyond the concerns of this book, we shall not develop the details of these different routes leading to chaos. Some excellent books have been written on this subject. We shall, however, provide some examples of simple systems and models showing how fractal structures enter into the description of chaotic structures.

A very common structure with very few (three) degrees of freedom which leads to a chaotic situation is the *simple forced pendulum*. A periodic force F cos ωt is exerted on this pendulum (Fig. 2.3.5). Its motion is governed by a

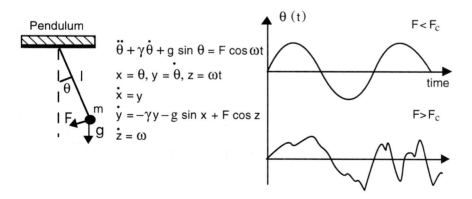

Fig. 2.3.5. Equations governing the simple forced pendulum (m=1, l =1). The system becomes chaotic when the applied force is greater than a certain critical threshold.

second order equation in θ, the angle of oscillation, which reduces to a system of three first order equations. The evolution of such a system may be studied by plotting what is known as a *phase portrait*, that is, each state of the system is represented by a point in phase space, the evolution then corresponding to a curve which does not intersect itself, and which starts from the point representing the initial state. In Fig. 2.3.5, we see that the simple forced pendulum's state is described by six coordinates ($\dot{x}$, $\dot{y}$, $\dot{z}$, x, y, z).

The simple forced pendulum is periodic in θ when the force applied is smaller than a certain threshold F_c and chaotic when the force F is greater than this threshold (Fig. 2.3.5). Many easily constructed systems display this type of behavior, such as a compass when it is placed in the alternating field of an electromagnet. We shall now describe in greater detail the *periodically struck rotator*, a rather artificial system with the pleasant property of being easily integrable via reduction to a simple two-dimensional discrete mapping, called a *first return mapping*.

The system consists of a rotator damped by friction (parameter γ), parameterized by an angle φ, and struck with period T, by an impulsive force whose intensity is a function[15] of this angle φ, namely K f(φ). A diagram of the model and the equations of motion are shown in Fig. 2.3.6.

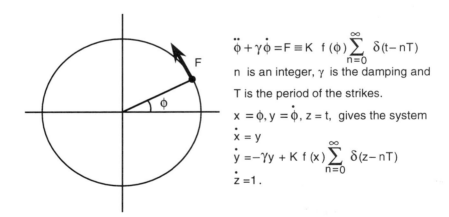

$$\ddot{\phi} + \gamma\dot{\phi} = F \equiv K \ f(\phi) \sum_{n=0}^{\infty} \delta(t - nT)$$

n is an integer, γ is the damping and

T is the period of the strikes.

$x = \phi, y = \dot{\phi}, z = t,$ gives the system

$$\dot{x} = y$$

$$\dot{y} = -\gamma y + K \ f(x) \sum_{n=0}^{\infty} \delta(z - nT)$$

$$\dot{z} = 1.$$

Fig. 2.3.6. Equations governing the periodically struck rotator.

After integration over a period T, where the limits are times just before a strike occurs, this fairly simple system may be reduced to a two-dimensional mapping for the pair of variables

$$(x_n, y_n) = \lim_{\varepsilon \to 0} \{x(nT - \varepsilon), \ y(nT - \varepsilon)\}.$$

---

[15] The function f is arbitrary but the nature of the chaos generated depends on its form (see the example on the next page of the first return mapping).

Thus, we find that (Schuster, 1984)

$$x_{n+1} = x_n + \frac{1-e^{-\gamma T}}{\gamma}\, [y_n + K\, f(x_n)]$$

$$y_{n+1} = e^{-\gamma T}\, [y_n + K\, f(x_n)]\,.$$

The evolution of the solutions of an equation of this type may be represented graphically as the variation of a point in the plane (x_n, y_n) as a function of time, that is of n, with the pair (x_0, y_0) as initial conditions.

The set of points (x_n, y_n) does not provide all of the system's phase portrait, but merely a two-dimensional section (associated to a two-dimensional mapping) composed of discrete points corresponding to photos (stroboscopic images) of the rotator taken at times separated by multiples of T. This sort of representation is known as a *Poincaré section*. It is the trace of the phase space motion passing through a subspace, which is easier to visualize.

First return maps or Poincaré maps

This is a mapping of the form

$$\boxed{\vec{x}_{n+1} = A\,(\vec{x}_n)}\,. \tag{2.3.-4}$$

The one-dimensional quadratic map or logistic map,

$$x_{n+1} = r\, x_n\, (1 - x_n), \tag{2.3.-5}$$

has, because of its extraordinarily rich nature, played a fundamental role in the study of processes leading to deterministic chaos. The discovery of its remarkable properties is due to Metropolis *et al.* (1973) and especially to M.J. Feigenbaum (1978). It is the prototype of mechanisms generating an infinity of subharmonic bifurcations (as r increases from 1 to 4). In this situation we speak of *subharmonic cascade*. The quadratic map is a special case of the Poincaré map of the periodically struck rotator, obtained in the limit of strong damping $(\gamma \to \infty)$, and as $K \to \infty$, with $\gamma/K = 1$, and f chosen such that $f(x) = (r - 1)x - r\, x^2$.

Subharmonic cascade

In the case of the quadratic map, for $1 \le r \le r_1 = 3$ (Fig. 2.3.7), x_n tends to a fixed point x* defined by x*= A(x*), that is, x*=1-1/r. In the case of the struck rotator, x_n is the angle of rotation just before the nth strike (given with periodicity T), x_{n+1} is the first return of the angle a time T later. At the end of a large enough number of strikes this angle stabilizes at a value $\phi = x^*$. Only the frequency $f_0 = 2\pi/T$ (and its harmonics) appears in the spectrum of the struck pendulum's motion in the limit case above. When r reaches the value r_1, a first bifurcation appears, and for $r_1 < r \le r_2$ (= $1+\sqrt{6} \cong 3.44$) x_n tends rapidly to a

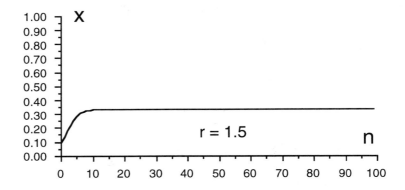

Fig. 2.3.7. Iteration of the quadratic map when the attractor reduces to a fixed point
$$x* (1 \leq r \leq r_1 = 3).$$

situation where it oscillates between two values, $x^{(1)}_1$ and $x^{(1)}_2$ defined by $x^{(1)}_i$ = $A[A(x^{(1)}_i)]$. Clearly $x*$ remains a solution to this last equation but is now unstable. If this result is applied to the struck rotator we can see that the impulse is delivered each time the angle ϕ takes the values $\phi^{(1)}_1$ and $\phi^{(1)}_2$. The frequencies f_0 and $f_0/2$ (and their harmonics) thus appear in the spectrum. Hence the term *subharmonic cascade*.

Then at $r = r_3$ a new bifucation appears and x_n tends towards a four point cycle (Fig. 2.3.8) (and the frequencies f_0, $f_0/2$, $f_0/4$ appear). After k bifurcations, a cycle of 2^k points $x^{(k)}_i$ is obtained with subharmonics ranging down to $f_0/2^k$. The set $x^{(k)}$ of the $x^{(k)}_i$, which form a cycle of 2^k points, is called an *attractor* as in this case the sequence of the x_n ends up converging to this set whatever the initial condition. In this way bifurcations continue to

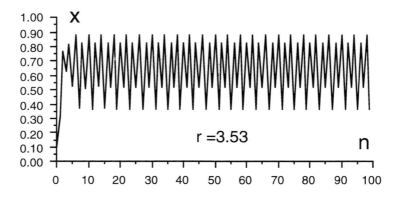

Fig. 2.3.8. Iteration of the quadratic map when the attractor is a cycle of four points.

follow until a value $r_\infty = 3.5699456...$ is reached. The limit set $x^{(\infty)}$ is now no longer periodic; it is then called a *strange attractor* and is a fractal. Fig. 2.3.10 represents the distribution of fixed points $x^{(k)}_i$. As r increases from 1 to r_∞, their structure resembles the successive iterations of a Cantor set. An important universal quantity is the limit of the ratio of distances between successive bifurcations, defined as

$$\delta = \lim_{k \to \infty} \frac{r_k - r_{k-1}}{r_{k+1} - r_k} = 4.669\ 201\ 609... \quad . \quad (2.3\text{-}6)$$

Finally, a chaotic situation appears for $r_\infty \le r \le 4$ (Fig. 2.3.9). The subharmonics $f_0/2^k$ then disappear one after the other (in the reverse order of their appearance) while a noise develops.

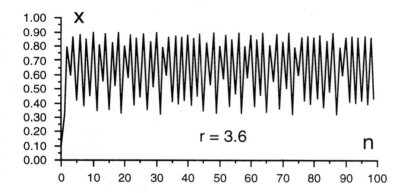

Fig. 2.3.9. *Iteration of the quadratic map when the attractor becomes aperiodic. This is a chaotic situation: the sequence of x_n is distributeb over [0,1] generating a fractal structure. The sequences depend very sensitively on the initial value.*

Another remarkable fact is that the domain $r_\infty \le r \le 4$ is not chaotic everywhere. There exist windows of periodicity which can be seen very clearly in Fig. 2.3.10. Some of them have periodicity 3 (e.g., when $r = 1 + \sqrt{8}$) with a new subharmonic cascade leading to chaos. For $r > 4$ there are no longer any stationary states, the sequence of the x_n is no longer bounded.[16]

The Hausdorff dimension of the set $x^{(\infty)}$ (associated with r_∞) has been numerically and analytically determined by Grassberger (1981): $D = 0.5388...$

Rayleigh–Bénard instability and subharmonic cascade

When a liquid with a positive coefficient of expansion is placed between two horizontal plates held at fixed temperatures, T for the upper plate and $T + \Delta T$ for the lower plate (Fig. 2.3.11), the liquid tends to move upwards from the bottom, producing an instability known as Rayleigh–Bénard instability.

---

[16] More precisely, only a bounded Cantor invariant remains.

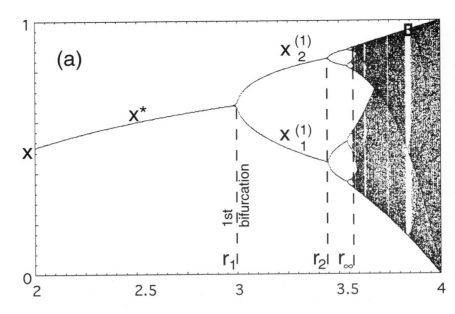

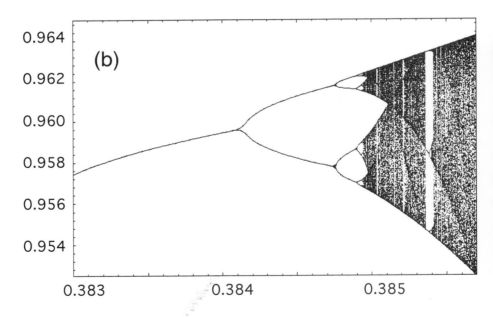

Fig.2.3.10. Attractor of the logistic map. Figure (a) shows the complete attractor
(except for the single fixed point range [1,2]). Figure (b) is the magnification of the
small rectangle in figure (a) inside a window of periodicity 3, displaying the self-
similarity.

For low values of ΔT energy transfer occurs simply through conduction without any motion on the part of the liquid, since the structure is still stable. Then for values of ΔT above a certain threshold (namely Ra > Ra$_c$ the relevant parameter being the Rayleigh number Ra[17]), convective rolls appear, with fixed spatial position, displaying a series of ascending and descending currents as shown in Fig. 2.3.11.

Fig. 2.3.11. Rayleigh–Bénard cell: a temperature difference induces a density difference in the fluid and also an instability in the presence of a gravitational field. Rolls appear which become unstable and chaotic beyond a critical ΔT.

Moreover, both directions of rotation of the rolls are equiprobable. At the threshold there is thus a first bifurcation between these two possible rotation states. As the temperature is increased the number of possible states also increases and the presence of so many degrees of freedom allows a turbulent situation to develop: the structure of the rolls is completely destroyed.

In an experiment carried out by Libchaber, Fauve and Laroche (1983), the liquid used is mercury. The experiment is performed at a very low Prandtl number (Pr = $v_0/D_T \cong 0.03$). The mercury is placed between two thick copper plates. As the cell is opaque the temperature fluctuations in the fluid are measured at a point (Fig. 2.3.12) using a bolometer. To stabilize the orientation of the rolls, the cell is placed in a constant magnetic field. By increasing ΔT, the Rayleigh number crosses the threshold Ra$_c$ (several precautions are necessary at this first stage of the experiment), then successive bifurcations appear, corresponding to frequencies $f_1/2$, $f_1/4$, $f_1/8$... The harmonics of odd rank are also present in the spectrum. The ratio of convergence of the bifurcations is found to have a value of 4.4, reasonably close to the universal theoretical value of 4.669...

One (very approximate) model of Rayleigh–Bénard convection was given by Lorenz (1963). It is, as with the previous examples, a system of 3 first order equations with second degree terms, xy and xz, obtained by simplifying as far as possible the Navier–Stokes and heat equations. By expanding these equations in spatial Fourier modes up to order 3 in the lowest mode, and ignoring the coupling between other modes, Lorenz's system of equations is obtained:

---

[17] Ra is directly proportional to ΔT, to the coefficient of volumic expansion of the fluid and to its weight in a cube of side equal to the distance between the plates, and inversely proportional to the viscosity μ and to the thermal diffusivity D_T.

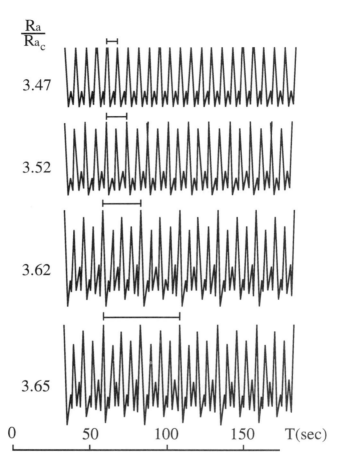

Fig. 2.3.12. Subharmonic cascade in a thermoconvection experiment (Libchaber et al. 1983). The segments show the length of a period of temperature fluctuation at a point in the fluid.

$$\dot{x} = -\sigma x + \sigma y ,$$
$$\dot{y} = -zx - r x - y ,$$
$$\dot{z} = xy - b z ,$$

(2.3-7)

where σ, r, and b are the parameters of the system; x measures the level of convective overturning, y and z measure the horizontal and vertical temperature variations, σ corresponds to the Prandtl number, r is proportional to the Rayleigh number, and reflects the fact that the horizontal and vertical temperature variations do not damp down at the same rate.

Iterated maps in the complex plane

Iterations in the complex plane form a very interesting class of mappings for mathematicians. They have given rise to graphical creations (thanks to the use of modern computers) of an until now unknown richness (see, e.g., Peitgen and Richter, 1984).

As an example, let us take the following mapping A of the complex plane:

$$z \rightarrow A(z) = z^2 + C \ . \tag{2.3-8}$$

The *Julia* and *Mandelbrot sets* of this mapping are defined by the iteration

$$z_{n+1} = z_n^2 + C \tag{2.3-9}$$

with the two fixed values, C and z_0, as the initial state.

To relate this to what we have just been discussing, Eq. (2.3-9) generalizes the equation of the logistic map (2.3-5) to the complex plane (up to an irrelevant translation in the coordinates). Notice that if we write

$$dx_n/dn \approx x_{n+1} - x_n = x_n^2 - x_n - y_n^2 + a \quad \text{or} \quad \dot{x} = x^2 - x - y^2 + a \ ,$$
$$\text{and} \quad d \ y_n/dn \approx y_{n+1} - y_n = x_n y_n - y_n + b \quad \text{or} \quad \dot{y} = xy - y + b \ ,$$

this iteration corresponds to a dynamic equation in the plane (where C = a+ib). However, the discrete equation represents a much more complex behavior than does the continuous one (which is constrained by the noncrossing condition).

Julia sets J_c

We fix C and look at the sequence $z_0 \rightarrow z_1 \rightarrow z_2$ ... For many values of z_0, $|z_n| \rightarrow \infty$ when $n \rightarrow \infty$. J_c is defined as the frontier of these diverging z_0. Points in the immediate neighborhood of the Julia set diverge towards infinity exponentially (exp λn).[18] To construct J_c it is easier to use the inverse mapping: $z_0 \rightarrow z_{-1} \rightarrow z_{-2}$ ..., where

$$z_{n-1} = \pm\sqrt{z_n - C} \tag{2.3-10}$$

This series converges exponentially to the Julia set. As there are two possible values for the inverse map, they are chosen at random with, say, equal probability.

J_c can also be thought of as the frontier of the set of points z_0 for which the z_n remain bounded, called the *basin of attraction of the map* (and not necessarily connected).[19] J_c depends sensitively on C. Some Julia sets are a single piece, others are only clouds of points (Cantor sets) (Fig. 2.3.13). The simplest case occurs when C = 0. The mapping $z_{n+1} = z_n^2$ starting from

---

[18] In situations relating to turbulence, the exponent λ is called the *Lyapounov exponent*. The development of a chaotic flow is more difficult to visualize, the greater the rapidity of the divergence of the trajectories. Hence the importance of knowing λ.

[19] Eq. (2.3-9) may therefore be iterated starting from points z_0 of the basin of attraction. However, if the latter is a Cantor set, it may prove difficult to find such points.

$z_0 = \rho\ e^{i\phi}$, gives $z_n = \rho^{2n}\ e^{2in\phi}$. If $\rho > 1$, $z_n \to \infty$, and if $\rho < 1$, $z_n \to 0$. The Julia set J_0 is the circle $|z| = 1$, and the basin of attraction is the interior of the circle.

On the subject of Julia sets, it is interesting to consult the papers of Julia and Fatou who managed to work out a number of characteristics of these sets without, of course, the use of computers. See also Brolin (1965) and, for more recent work, Blanchard (1984).

Mandelbrot   set

This is the set $\mathcal{M}$ of those C such that for $z_0 = 0$, z_n remains bounded (Mandelbrot 1980) (Fig. 2.3.13). On a deeper level, $\mathcal{M}$ is also the set of those C such that J_c is connected. The speed of divergence outside $\mathcal{M}$ is related to solutions of the Laplace equation, $\mathcal{M}$ being thought of as a charged conductor and the speed of divergence as proportional to the electric field.

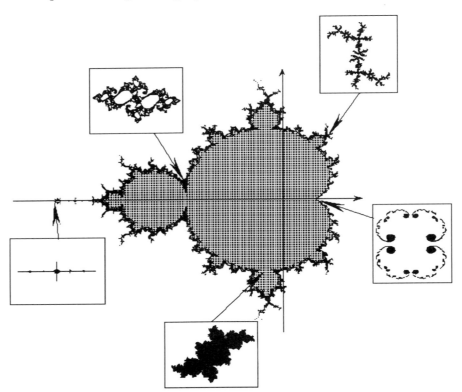

Fig. 2.3.13. Mandelbrot set (shaded grey) for the iteration $z \to z^2 + C$. In the insets the Julia sets correspond to the values of C indicated by arrows shown. The Mandelbrot set is surrounded by a black border (like a beach on the seashore), which gives an idea of the speed with which the values z move away from the set during iteration (the narrower the beach the faster the point moves away).

All these different sets have a fractal structure. Fig. 2.3.14 represents the correspondence between the Mandelbrot set in the complex plane and the bifurcated structure of the modified quadratic first return mapping (logistic map) $x_{n+1} = x_n^2 + C$ along the real axis (after Peitgen and Richter, 1984, p. 11).

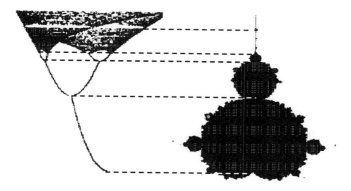

Fig. 2.3.14. *Correspondence between the Mandelbrot set land the logistic map $x_{n+1} = x_n^2 + C$. Notice the repetition of the cactus motif at each bifurcation.*

Strange attractor

We have seen that periodic attractors correspond to perfectly predictable signals, in the sense that knowledge of a sufficiently long sequence of the signal [$\theta(t)$ for example] enables us to predict the whole of the future signal with an accuracy equal to that of the known sequence. This is no longer the case when the attractor becomes aperiodic, since it is impossible in practice, whatever the portion of signal considered, to predict the future signal beyond a very limited time because of the *extreme sensitivity to initial conditions*. The histories of two points starting very close to one another have nothing in

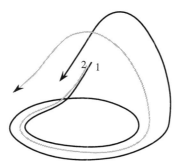

Fig. 2.3.15. *Schematic representation of two trajectories (1) and (2) initially very close, moving apart exponentially (positive Lyapounov exponent) then returning towards the spiral (after Abraham and Shaw 1983, part 2).*

common at the end of a large enough finite time. The strange situation then occurs where the histories (trajectories in phase space) diverge exponentially,[20] but where there is still an attractor. For topological reasons, this is not possible in just any phase space. Since two trajectories cannot intersect each other,[21] a space of at least three dimensions is required. It is possible for trajectories to remain confined to a bounded domain while still possessing a sensitivity to initial conditions which causes them to diverge exponentially at first (as Fig. 2.3.15 shows). This again is a *strange attractor*.

The Hénon attractor

Hénon (1976) proposed a simpler set of Lorenz equations, obtained by discretizing time, and thereby transforming Lorentz's system of three differential equations into a two-dimensional mapping. He studied the following mapping of the plane into itself:

$$X_{k+1} = Y_k + 1 - \alpha (X_k)^2$$
$$Y_{k+1} = \beta X_k .$$
                                                                    (2.3-11)

This type of iterated application lends itself particularly well to a throrough investigation. Whatever the initial points chosen within the application's basin of attraction, the successive iterations converge very rapidly towards a strange attractor, the Hénon attractor, whose Hausdorff dimension is $D = 1.26...$ Moreover, it is easy to verify by calculating the Jacobian of the transformation (2.3-11) that the areas are multiplied at each iteration by a certain factor.

Multifractality of strange attractors

Multifractal analysis is a tool well adapted to characterizing strange attractors. Here is an example taken from an experiment. In 1985, Jensen and his colleagues studied the thermoconvection of mercury in small parallelepiped cells (7 mm x 14 mm by 7 mm high). In the chosen geometric configuration the ratio between the height and width of the Rayleigh–Bénard cell forces the number of rolls to be limited to two. The number of degrees of freedom is thus reduced and only secondary instabilities appear. As ΔT is increased, the rolls break down at a frequency f_1 (the Fourier spectrum containing f_1, $2f_1$,...) and the system's state may be represented by a point on a circle, turning (nonuniformly) with frequency f_1; as ΔT is increased further a new bifurcation arises at frequency f_2 (the Fourier spectrum then contains f_1, $f_1 - f_2$, $2f_2$,...) and the system becomes quasi-periodic, since there is no reason that f_1 and f_2 should be commensurable.

---

[20] It should of course be noted that this exponential divergence of initially very close trajectories occurs only during a limited interval of time (see Fig. 2.3.15).
[21] From the point of intersection I there would be two paths (with the same initial conditions I) which is impossible.

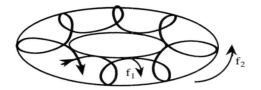

Fig. 2.3.16. *Representation by a point on a torus T^2 of the state of a system with two frequencies. If the frequencies are incommensurable, the image of the trajectory never repeats itself in the course of time; if f_1/f_2 is a rational fraction, it repeats itself after a finite number of turns.*

Its state can be represented as a point on a torus[22] (Fig. 2.3.16).

In fact, this quasiperiodic situation does not endure, but is instead seen to fix on a frequency f_0 such that $f_0 = f_1 = 2f_2$, by self-adjustment[23] of the frequencies f_1 and f_2. In these systems a chaotic situation arises after only a few bifurcations, according to a scenario named after Ruelle, Takens and Newhouse, as the small size of the cell prevents a subharmonic cascade from developing. But as we saw above at least three dimensions are required to construct a strange attractor, so that with two frequencies, f_1 and f_2, the trajectories will never exponentially diverge, nor will chaos become established. If a third frequency appears, the trajectories now lie on a torus T^3 and chaos is accessible, the trajectories no longer being confined to a surface (as in Fig. 2.3.15).

The transition to chaos may occur directly from the torus T^2 without a third frequency emerging; a more general third degree of freedom will suffice[24] (e.g., due to an external field forcing the system). Because of this new parameter the trajectories leave the surface of the torus. The threshold for the appearance of chaos for which the torus T^2 "explodes" corresponds, as can be shown in the theoretic model of the Rayleigh-Bénard experiment, to the ratio $f_1/f_2 = (\sqrt{5}+1)/2$, the golden ratio.

In the Jensen et al. experiment (1985), the Rayleigh-Bénard system is forced by an electromagnetic oscillator: an alternating current of frequency f_{ext} is passed through the mercury, in the presence of a magnetic field parallel to the rolls, in such a way that vertical eddies are generated. These are nonlinearly coupled to the eddies which we have just described above (two convective rolls are present in the cell). The intensity of the alternating current allows the nonlinear coupling to be adjusted. With this perturbation a chaotic situation can be generated from two frequencies (f_1 and f_{ext} chosen to be in the golden ratio, $f_1 \cong 230$ mHz). In this case the chaos is described by a prechaotic distribution of trajectories on the torus, which still exists as thecritical threshold has only

---

[22] The trajectories would not intersect on the torus if it were unfolded, as a point on the torus is only defined up to multiples of 2π. The rule of no intersection is therefore not violated.
[23] This adjustment is also present in the experiment of Libchaber et al.
[24] The model of Curry and York (1977) is of this type.

Fig. 2.3.17. The two-dimensional attractor above is composed of 2500 experimental points. Theoretically the points are distributed on a closed curve, the spreading that we see here is due to the drifting of the experimental parameters. The variations in the density of points comes partly from projecting the three-dimensional curve onto the plane (Jensen et al. 1985).

just been reached. When this critical situation is established (Ra/Ra$_c$ = 4.09, giving f$_1$ a large amplitude), measurements of the temperature T are taken at a point in the fluid at periodic intervals (i.e., the points of the trajectories in a section of the torus are examined). The graph of T(t+1) against T(t) is shown in Fig. 2.3.17. To avoid effects of projection, the multifractal analysis was, in fact, carried out in a three-dimensional representation [T(t+2), T(t+1), T(t)]. The multifractal measure of the Poincaré section (Fig. 2.3.17) is obtained by following a method very similar to the one described in Sec.1.6.2. More precisely, to each volume element of radius ε, the neighborhood of a point $\vec{x}$ of this section, is associated a measure which is the probability of returning to this element after a fixed time. In practice, the number of steps m($\varepsilon,\vec{x}$) to return to the neighborhood of $\vec{x}$, over a fixed interval of time are counted, the probability then being $\mu(\varepsilon,\vec{x})$ = 1/m($\varepsilon,\vec{x}$). The multifractal measure is then found by

$$\langle \mu(\varepsilon)^{q-1} \rangle = \sum_{i \in boxes} \mu(\varepsilon,\vec{x}_i)^q = \varepsilon^d \sum_{\vec{x}} \mu(\varepsilon,\vec{x})^{q-1} = M_q(\varepsilon). \qquad (2.3\text{-}12)$$

As the summation ranges over all the points x, and not the boxes i as required by the box-counting method, a factor $\varepsilon^d/\mu(\varepsilon,\vec{x})$ must be included. Having calculated M$_q$(ε) by digitalizing the experimental attractor, τ(q) and hence f(α) are determined by Legendre transformation (Secs.1.6.2 and 1.6.4). The experimental results are shown in Fig. 2.3.18, along with the theoretic curve obtained by calculating the distribution in a section (a circle) of the torus, for the critical value of the frequency ratio (the golden number). In this way, there are no adjustable parameters and the comparison between theory and experiment proves to be excellent.

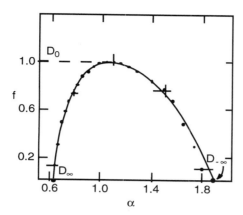

Fig. 2.3.18. Theoretical and experimental f(α) curves in the forced Rayleigh-Bénard experiment. Comparison between theory (without adjustable parameters) and experiment is remarkable (Jensen et al. 1985).

We can see on this curve $f(\alpha)$ that its maximum D_0 is equal to one, as we would expect, since the support of the multifractal measure is a circle (or is topologically equivalent to a circle), a section of the torus T^2 of Fig. 2.3.16.

This represents one of the few cases where it proves possible to compare an experiment with the theoretical model of the multifractality of strange attractors.

CHAPTER 3

Natural Fractal Structures
...to the microscopic

This chapter will discuss fractal structures encountered in the physics of condensed matter and solids, that is, disordered media, thin evaporated films, porous media, diffused structures, colloids and aerosols, aggregates, deposited layers, polymers, and membranes. Aside from the interest arising from the description and understanding of the physical processes generating such peculiar structures, knowledge of the geometry of these structures may have important consequences in various technological fields: microelectronics, for disordered materials and thin or deposited films; petroleum recovery, for porous materials; gelification and rheological properties for polymers, etc. As in Chap. 2, physical phenomena of the same type (of the same universality class) will also be discussed.

3.1 Disordered media

Disordered media is the term used to denote media composed of a random agglomeration of at least two types of material. Examples include alloys, discontinuous deposits of metallic films, diluted magnetic materials, polymer gels, and general composite materials. This section will present percolation as a theoretical model, and describe experimental studies on evaporated films which, due to their two-dimensional nature, lend themselves to analysis. The study of other disordered materials such as alloys or composites is found in Chap. 5 on transport in heterogeneous media.

3.1.1 A model: percolation

Percolation plays a fundamental role in a considerable number of physical phenomena in which disorder is present within the medium. Examples include: heterogeneous conductors, diffusion in disordered media, forest fires, etc. In Table I below, extracted from Zallen (1983), we give a partial list of its applications. Here we can see the extent of the range of applications of

TABLE I: Areas of application of percolation theory

Phenomenon or system	Transition
Random star formation in spiral galaxies	: Nonpropagation / Propagation
Spread of a disease among a population	: Contagion contained/ Epidemic
Communications network or resistor network	: Disconnected / Connected
Variable distance jump (amorphous semiconductors)	: Analoguous to resistors networks
Flow of a liquid in a porous medium	: Localized / Extensive wetting
Composite materials, conductor – insulator	: Insulator / Metal
Discontinuous metallic films	: Insulator / Metal
Dispersion of metallic atoms in conductors	: Insulator / Metal
Composite materials, superconductor – metal	: Metal / Superconductor
Thin films of helium on surfaces	: Normal / Superfluid
Diluted magnetic materials	: Para / Ferromagnetic
Polymer gels, vulcanization	: Liquid / Gel
Glass transition	: Liquid / Glass
Mobility threshold in amorphous semiconductors	: Localized / Extended states
Quarks in nuclear matter	: Confinement / Nonconfinement

percolation; like fractal structures themselves, percolation is found in models of galaxy formation as well as in the distribution of quarks in nuclear matter. The table reveals that the physicalsystems involved all undergo transitions. Within the framework of percolation, this transition is related to a phase transition. Indeed, the close parallel between percolation and phase transition is mathematically demonstrable.

Many experiments bring to light the characteristics of percolation. We shall describe a certain number in this section, but more will appear elsewhere in this book during discussion of other topics.

What is percolation?

Percolation was introduced in an article published by Broadbent and Hammersley in 1957. The word "percolation" originated from an analogy with a fluid (the water) crossing a porous medium (the coffee). As it transpires, the invasion of a porous medium by a fluid is in fact more complex (see Sec. 3.2) than the theoretical model that we are going to present now.

By way of a simple introduction to percolation, imagine a large American city whose streets form a grid (square network) and a wanted person, "Ariane", who wishes to cross the city from East to West (Fig. 3.1.1). Now, the police in this city are badly trained, choosing to set up roadblocks in the streets merely at random. If p_B is the proportion of streets remaining open, the probability of Ariane crossing the city decreases with p_B, and the police notice

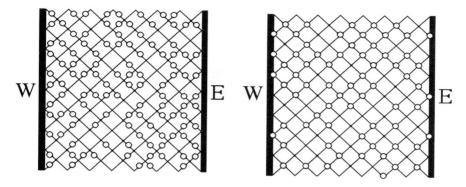

Fig. 3.1.1. On the left, bond percolation: out of 224 bonds, 113 are broken (white points), however there is still a route from East to West. On the right, site percolation: out of 128 sites, 69 are blocked, but there is no route through in this example.

(the city is very large) that with $p_B = p_{Bc} = 1/2$ they are almost certain (in the probabilistic sense) of stopping Ariane. The problem that we have just devised is one of bond percolation (the streets which form the bonds between crossroads are forbidden). The concentration p_{Bc} is called the bond percolation *threshold*.

For this result the city must be very large, as this threshold, p_{Bc}, is in fact a limit as the size (of the city, network, etc.) tends towards infinity. We shall return to the subject of the effects of finite size later.

But let us return to Ariane as, after some thought, the police, although continuing to set up roadblocks at random, have now decided to place them at crossroads. They now notice that the number of roadblocks required is smaller. There are two reasons for this: first, the number of crossroads is half that of the streets (if the city is large enough to ignore the outskirts), and second, if p_S is the proportion of crossroads remaining empty, we find a percolation threshold $p_{Sc} = 0.5927...$, i.e. it is sufficient to control about 41% of the crossroads. This is a case of site percolation (since in a crystalline network the "crossroads" are the sites occupied by the atoms).

If our network of streets is replaced by a resistor network some of which are cut, the current passing between W and E as a function of the proportion of resistors left intact is given by the graph below (Fig. 3.1.2). The current is zero for $p < p_c$, while, over the percolation threshold p_c it increases. We shall study this in greater detail in Sec. 5.2, which focuses on transport phenomena.

To summarize, we can say that the percolation threshold corresponds to the concentration (of closed bonds, etc.) at which a connected component of infinite extent appears (or disappears): a path is formed and, in materials where connected components conduct, the current flows. We can now begin to see why transitions were mentioned in the table shown at the start of the section.

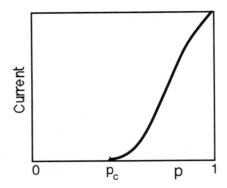

Fig. 3.1.2. Variation in the current passing between two opposite faces of a network of resistors of which a proportion (1–p) are cut at random. There is a threshold p_c of intact resistors such that a path allowing current to flow can be found. This threshold is better defined the larger the network.

The graph shown in Figure 3.1.2 could also represent, say, the conductivity of a random distribution of insulating and conducting balls (p_c is then about 0.3).

Percolation over regular networks

The first studies on percolation were for regular networks (relating to crystalline structures, which are mathematically simpler). Here we shall outline some properties and define various substructures which will of use further on. For a more precise and mathematical treatment consult Essam, 1972 & 1980 and Stauffer, 1985.

Except in certain two-dimensional networks (indicated by an asterisk in Table II), where, by symmetry, p_c can be calculated exactly, the percolation threshold can only be found by numerical simulation or by series extrapolation (Table II). The method employed in calculating the percolation threshold by

Table II: Percolation thresholds for some common networks

p_c	Site (p_{Sc})	Bond (p_{Bc})
Honeycomb	0.6962	0.65271*
Square	0.59275	0.50000
Triangular	0.50000*	0.34719
Diamond	0.428	0.388
Simple cubic	0.3117	0.1492
Base centered cubic	0.245	0.1785
Face centered cubic	0.198	0.199

numerical simulation consists in finding the concentration at which a percolation cluster appears in a network of size L^d, and then extrapolating the results as $L \rightarrow \infty$. This method is not very precise in practice, and, with enough memory and computer time, more elaborate calculational techniques on stripes or concentration gradients are preferable. Calculating the threshold by series extrapolation involves analytically determining the probabilities of occurrence of as many as possible finite cluster configurations [see for instance the calculation of $P_\infty(p)$, Eq. (3.1-4), where the clusters used in the calculation are only the very smallest ones], and then trying to extrapolate to very large clusters. Note that in two-dimensions there are duality transformations between networks (e.g., the transformation star $\Leftrightarrow$ triangle) which induce relationships between their respective thresholds, thus enabling exact calculations to be made in certain cases.

In the case of a *Cayley tree*, in which each site has z first neighbors (also known as a Bethe network), the exact solution may also be found: $p_c = 1/(z-1)$ (see below for the mean field approach). In Fig. 3.1.3, a part of a Cayley tree is shown with $z = 3$:

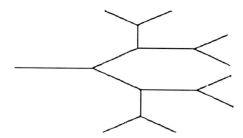

Fig. 3.1.3. Part of a Cayley tree with z = 3 branches attached to each site (the
network continues indefinitely, of course, along all of its branches).

Subset of occupied sites

Below the threshold, the percolation network is composed solely of finite clusters, a cluster being defined as a set of connected sites or bonds. If s is the size of a finite cluster (on the diagram below two finite clusters have been indicated: $s = 1$ and $s = 4$) and if n_s is the number of clusters of size s per site, sn_s is the probability of finding a given site belonging to a cluster of size s. Since below the threshold an occupied site is necessarily in a finite cluster, the probability of a site being occupied leads to the relation

$$\Sigma_s \, s \, n_s = p.$$

As the concentration reaches the percolation threshold, i.e., when $p \geq p_c$ ($\cong 0.59$ for a square lattice), an *infinite cluster*, or *percolation cluster*, appears (shaded gray in Fig. 3.1.4). At the threshold itself, the probability, P_∞, of a

p = 0.35   p = 0.59   p = 0.87

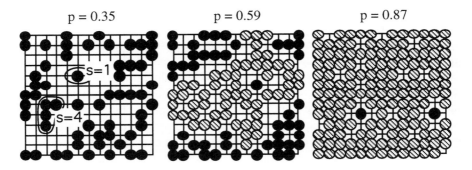

Fig. 3.1.4. Square network (L × L, L = 13) for three concentrations, p, of occupied sites. For sufficiently small p there are only finite clusters (s = 1,....,4,...). At a certain concentration p$_c$ (L), a percolation cluster (shaded grey) emerges. At higher concentrations this cluster invades the network. As L→ ∞, p$_c$ (L)→ p$_c$ and the percolation cluster becomes an infinite fractal cluster.

site to belong to this cluster is still zero. This infinite cluster which emerges at the percolation threshold is often called the *incipient infinite cluster*. Above the threshold, the probability P$_\infty$ grows very rapidly. An occupied site is now either in a finite cluster or in the infinite cluster, and hence

$$\boxed{P_\infty + \sum_s s\, n_s = p}$$ (3.1-1)

In the neighborhood of p$_c$ the function P$_\infty$(p) behaves as

$$P_\infty(p) \propto (p - p_c)^\beta$$ (3.1-2)

with $\beta = 5/36$ for two-dimensional networks (d = 2)
and $\beta \cong 0.44$ for three-dimensional networks (d = 3).

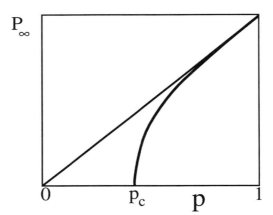

Fig. 3.1.5. Probability of a site being in the infinite cluster. Away from the threshold, when nearly all the sites belong to this cluster, P$_\infty$ varies like p.

We shall frequently observe this kind of behavior on the part of physical quantities. The exponents (here denoted β) are numbers which, generally speaking, depend only on the model being studied (in this case, percolation) and the Euclidean dimension d of the space. They are, for example, independent of the shape of the network (square, triangular, or honeycomb). In particular, β is the same for site and bond percolation. There is said to be *universality* present. This characteristic is also found in the case of phase transitions close to the critical temperature T_c, including magnetism close to the Curie temperature.

Some of these exponents represent *fractal dimensions*, although the *mathematical* reasons are as yet unknown.

Calculating $P_\infty(p)$

We can, in theory, calculate the probability $P_\infty(p)$ arithmetically by considering the following equation[1]

$$P_\infty(p) = 1 - P_{finite}(p) \qquad (3.1\text{-}3)$$

and calculating $P_{finite}(p) = \Sigma_{s=0}^{\infty} P_s(p)$, where P_s is the probability for a site to belong to a cluster of size s. For site percolation over a square network (an empty site is represented by a white disk, an occupied site by a black disk), we have, for example,

$$P_0 = \qquad \bigcirc \qquad = q \equiv 1 - p$$

$$P_1 = \qquad \begin{matrix} \bigcirc \\ \bigcirc \ \bullet \ \bigcirc \\ \bigcirc \end{matrix} \qquad = p\,q^4$$

$$P_2 = \qquad \begin{matrix} \bigcirc \\ \bigcirc \ \bullet \ \bigcirc \\ \bigcirc \ \bullet \ \bigcirc \\ \bigcirc \end{matrix} \qquad = 2 \times 4 \ p^2 q^6$$

... and in general (s > 0),

$$P_s = s\,n_s = \Sigma_t \, s \, g_{st} \, p^s \, q^t , \qquad (3.1\text{-}4)$$

where t is the size of the perimeter of the cluster, i.e., the set of empty sites which are first neighbors to one of the sites of the cluster. The difficulty, of course, is in calculating g_{st} : no expression is known for large s, and yet all the properties of percolation arise from the statistics of large clusters. Series of this type have been studied extensively, especially when the first detailed work on percolation was being carried out (de Gennes et al., 1959; Essam, Sykes, and

---

[1] The probability of being in either a finite (s ≥ 0) or infinite cluster is 1.

Essam: see Essam, 1980). They provided the first values of p_c, the critical exponents, etc.

Mean field approach

Failing a rigorous method of determining the probability P_∞ (p), an approach based on a *mean field* (also known as an *effective field*) calculation has been tried. The first studies of this kind are due to Flory and Stockmayer (see Flory, 1971) who, in as early as 1941, were investigating the appearance of gelification in polymers by developing a mean field approach over a Bethe network, before the idea of percolation had even emerged (we shall discuss this again later in Sec. 3.5 on gels).

Mean field calculations work as follows: taking Fig. 3.1.6(a) as our starting point, we see that for site **a** to belong to a finite cluster it must either be an empty cluster, with probability 1–p, or be occupied, with probability p, and then connected to a finite cluster by one of its z neighbors. So we can now write

$$P_{finite}(p) = (1-p) + p\,G^z\,, \qquad (3.1\text{-}5)$$

where G is the probability that a site neighboring **a** belongs to a finite cluster $(0 \leq G \leq 1)$. In this equation we must assume independence between the branches (no loops).

Taking one of these z neighbors, we see that it is itself linked to (z–1) second neighbors and so we have,

$$G(p) = (1-p) + p\,G^{z-1}\,. \qquad (3.1\text{-}6)$$

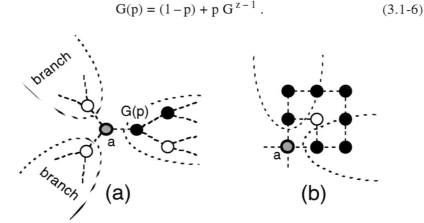

*Fig. 3.1.6. (a) Diagram showing mean field approximation. Site **a** of a finite cluster may either be occupied or empty; if it is occupied it must have each of its z (= 3) neighboring sites belonging to a finite cluster, which they each do with probability G. These sites must in turn either be empty or connected to one of the z–1 second neighbor sites, again belonging to finite clusters with probability G. (b) This approach is only approximate if there are loops linking the branches.*

This approach does in fact completely resolve the problem of percolation over a Bethe network (which has no loops).

i) For $z = 2$, the network is a linear chain; so Eq. (3.1-6) is now

$$G = 1 - p + pG. \qquad (3.1\text{-}6a)$$

Hence, when $p \neq 1$, Eq. (3.1-6a) has as its solution $G \equiv 1$, and so [from (3.1-5)]: $P_\infty \equiv 0$. When $p = 1$, G is not determined by Eq. (3.1-6a), but as all the sites of the chain are occupied, $P_\infty = 1$ and the percolation threshold occurs at $p_c = 1$.

ii) For $z = 3$, the network is hexagonal (or honeycomb); So Eq. (3.1-6) now becomes:

$$G = 1 - p + pG^2. \qquad (3.1\text{-}6b)$$

This equation possesses two roots: $G = 1$ and $G = (1{-}p)/p$. The solution $G = (1 - p)/p$ is only valid for $p \geq p_c = 1/2$, as we require $0 \leq G \leq 1$, and in this case it may be shown that $G = 1$ is not a valid solution. Hence,

(a) when $p < p_c = 1/2$, there is only one solution: $G = 1$ and $P_\infty = 0$,

(b) when $p > p_c = 1/2$, $G = (1 - p)/p$ is the unique solution and an *infinite cluster* occurs, with the probability of a site belonging to it given by

$$P_\infty = p - q^3/p^2.$$

In the neighborhood of p_c ($p = 1/2 + \varepsilon$) we find:

$$G \cong 1 - 4\varepsilon \quad \text{hence} \quad P_\infty(p) \cong 6\varepsilon = 6(p - p_c).$$

iii) Similarly, it can be shown for any value of z that

$$P_\infty(p) = 2z/(z-2) \ (p - 1/(z-1))^\beta + \dots \text{ with } \beta = 1. \qquad (3.1\text{-}7)$$

Thus, a *mean field* approach yields $\beta = 1$.

Eqs. (3.1-5) and (3.1-6) provide exact results for Bethe networks, hence much interest has been shown in this type of network. They represent textbook examples of mean field approaches to percolation problems.

It is interesting to compare the first iterations in mean field calculations with those of an exact calculation. Thus for $z = 4$ (square network), although we find the beginning of the series for P_{finite} to be the same up to clusters of size two, from there they differ. In mean field,

$$P_{finite}(p) = q + p \, q^4 + 2 \times 4p^2q^6 + 3 \times 18p^3q^8 + 4 \times 88p^4q^{10} + \dots , \qquad (3.1\text{-}8a)$$

whereas the exact calculation gives,

$$P_{finite}(p) = q + p \, q^4 + 2 \times 4p^2q^6 + 3 \times (12p^3q^7 + 6p^3q^8) +$$

$$4 \times (36p^4q^8 + 32p^4q^{10} + 8p^4q^{10}) + \dots . \qquad (3.1\text{-}8b)$$

The two series differ as soon as loops are possible, or to be more precise, as soon as sites on the perimeter have several first neighbors in the cluster (loops including a perimeter site; this occurs for clusters of size 3).

Summary of three things to remember

(1) The mean field approach is equivalent to solving for a Bethe network, that is, a network in which loops never occur. It is the loops which make the problem insoluble at present by introducing correlations between the probabilities.

(2) The mean field approach provides a set of critical exponents which may be completely determined. We have calculated β (= 1) here. The mean field exponents are themselves *universal.* Their values are shown in the table at the end of the section under the heading "Bethe".

(3) The dimension at which the mean field becomes exact is called the *critical dimension*, which is when the probability of loops forming is zero. If the dimension of the space is increased, the room available for branches in a percolation cluster grows sufficiently large to make the occurrence of loops very rare (i.e., of probability going to zero as the size of the incipient infinite cluster goes to infinity). Consequently, for all dimensions $d \geq d_c$, the mean field becomes exact. This dimension is mainly of theoretical interest [because it can be expanded in the neighborhood of d_c: $(d = d_c - \varepsilon)$ it can be compared with other methods]. In the case of percolation, the critical dimension $d_c = 6$ (and the fractal dimension of the infinite cluster is then $D = 4$, see below).

Besides $P_\infty(p)$, there are other quantities of great physical significance. These include:

— $S(p)$: the mean number of sites in a finite cluster (Fig. 3.1.7).

— $\xi(p)$: the connection length, mean radius of the finite clusters.

To these we might add:

— $\Sigma(p)$: the mean macroscopic conductivity in metal–conductor mixtures, about which we shall only say a few words in this section.

The mean size of the finite clusters, i.e., the mean number of sites contained in one of them, $S(p)$, diverges at p_c because of the emergence of an infinite cluster. By definition,

$$S(p) = \langle s \rangle = \Sigma_s \, s(sn_s)/\Sigma_s \, sn_s = \Sigma \, s^2 n_s \, /p \, . \qquad (3.1\text{-}9)$$

It behaves as

$$\boxed{S(p) \propto | \, p - p_c \, |^{-\gamma}} \qquad (3.1\text{-}10)$$

where $\qquad \gamma = 43/18$ when $d = 2.$
$\qquad\qquad\quad \gamma \cong 1.8 \quad$ when $d = 3.$

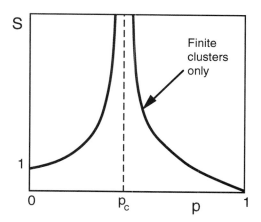

Fig. 3.1.7. Mean size of finite percolation clusters. S(p)'s behavior is governed by the same exponent above and below the threshold.

Remark: When $p > p_c$ the infinite cluster is, of course, not taken into account in the calculation of S(p).

Cluster radius and fractal dimension

A cluster of size s has a certain mean radius R_s. In practice, it is useful to define it by what is known as the *radius of gyration*, that is, the radius at which all the "mass" would need to be situated for there to be an equal moment of inertia. We let

$$R_s^2 = \frac{1}{s} \sum_{i=1}^{s} |\vec{r}_i - \vec{r}_0|^2, \qquad (3.1\text{-}11a)$$

where $\vec{r}_0 = \frac{1}{s} \sum_{i=1}^{s} \vec{r}_i$ (center of mass), i being a site of the cluster.

This may also be written

$$R_s^2 = \frac{1}{2s^2} \sum_{i,j} |\vec{r}_i - \vec{r}_j|^2. \qquad (3.1\text{-}11b)$$

The correlation (or, better here, connection) length $\xi(p)$ is the mean radius of gyration of *all* the finite clusters. It gives an idea of the average distance at which the connectivity makes itself felt. Since there is a probability sn_s of a randomly chosen site, i, belonging to a cluster of size s, and there are s sites j corresponding to such a site, we define ξ^2 by the relation

$$\xi^2(p) = \langle R_s^2 \rangle = \frac{2 \sum_s R_s^2 \, s^2 \, n_s}{\sum_s s^2 \, n_s}. \qquad (3.1\text{-}12)$$

As p approaches p_c from below, an infinite cluster emerges, that is, one of infinite radius of gyration. Thus, ξ must diverge at p_c. When $p > p_c$, ξ is still defined as the average over finite clusters; it diverges as p tends to p_c from above. As with the other quantities described earlier, ξ behaves according to a power law close to p_c. We may thus write (the notation v is adopted for the critical behavior of the correlation length whatever the type of model)

$$\xi(p) \propto |p - p_c|^{-v} \quad . \tag{3.1-13}$$

The exponent v is the same on either side of p_c. The values of v are

$$v = 4/3 \text{ when } d = 2; \quad v \cong 0.88... \text{ when } d = 3.$$

This correlation length is directly related to the correlation function, g(r), the probability that a site at distance r from an occupied site belongs to the *same finite cluster* [g(0) is normalized to one].

$$g(r) = \frac{\langle m(\vec{r_0}) \, m(\vec{r_0} + \vec{r}) \rangle}{\langle m(\vec{r_0}) \rangle} = \langle m(\vec{r_0}) \, m(\vec{r_0} + \vec{r}) \rangle / p \quad . \tag{3.1-14}$$

As the only contributions to g(r) come from occupied sites r_0 and r_0+r, $[m(r_0) = m(r_0+r) = 1]$, the average number of sites to which an occupied site r_0 is connected is $p \sum_r g(r)$, the summation ranging over all the sites of the network. This mean number is therefore $\sum s^2 n_s$, so that

$$p \, S(p) = \sum_s s^2 n_s = p \sum_r g(r) \quad . \tag{3.1-15}$$

The mean distance ξ is the weighted sum of the mean distances between two sites belonging to the same cluster, so we have

$$\xi^2 = \sum_r r^2 g(r) \Big/ \sum_r g(r) \quad . \tag{3.1-16}$$

The relation (3.1-13) enables us to evaluate the effects of finite size on the value of the percolation threshold. Indeed, for a sample of size L, the concentration at which the finite clusters percolate is on average such that $\xi(p) \cong L$. The apparent threshold is therefore

$$p_c^* \cong p_c - C \, L^{-1/v}. \tag{3.1-17}$$

Fractal dimension

By calculating quantities such as R_s or g(r), various relationships may be found concerning the distribution of occupied sites in a finite cluster, in the case of R_s, and their mean distribution, in the case of g(r). This allows the following question to be posed: are the size s and the radius R_s of a cluster related by a power law defining a fractal dimension?

Numerical experiments and phase transition considerations (renormalization group) demonstrate that this is precisely the case for large clusters at the percolation threshold. Indeed, the mass–radius relation may be written

$$s \propto R_s{}^D \qquad (p = p_c , s \to \infty). \qquad (3.1\text{-}18)$$

(At $p = p_c$, there exists an infinite cluster and many finite clusters whose structures merge with the infinite cluster as $s \to \infty$.)

   This fractal structure is also obtained by averaging over all the finite clusters: if we calculate the mean mass at $p = p_c$ of occupied sites belonging to the same finite cluster by integrating the correlation function g over an annulus between r and r+dr, we find

$$M(R) \propto \int_0^R r^{d-1} \, g(r) \, dr \propto R^D. \qquad (3.1\text{-}19)$$

This forces the asymptotic behavior of $g(r)$ to be

$$g(r) \propto \frac{1}{r^{d-D}} \qquad (p = p_c) . \qquad (3.1\text{-}20)$$

It can be shown that
$$D = 91/48 \cong 1.89 \quad \text{when } d = 2,$$
$$D \cong 2.5 \quad \text{when } d = 3.$$

When we no longer have $p = p_c$, the scaling function takes the general form

$$\boxed{g(r,\xi) \propto \frac{1}{r^{d-D}} \, \Gamma\!\left(\frac{r}{\xi(p)}\right)} \qquad (3.1\text{-}21)$$

and the mass in a ball of radius R is

$$\boxed{M(R,\xi) \propto \xi^D \Phi\!\left(\frac{R}{\xi(p)}\right) \quad \text{with } \Phi(x) = \int_0^x u^{D-1} \, \Gamma(u) \, du} \qquad (3.1\text{-}22)$$

These important expressions are worth discussing further. When $p = p_c$, ξ is infinite and g is a simple power law of the form (3.1-20), characteristic of a fractal distribution. But if $p \neq p_c$, the finite clusters have a mean radius ξ which is then the only significant distance in the system; there is of course a minimum distance a (distance between sites, for example), but when the clusters are sufficiently large this length plays no role. In general, there is also a maximum length, L (size of the sample), which is assumed here to be very large in relation to ξ. So we postulate that the ratio $g(r, \xi)/g(r,\infty)$ is a dimensionless scaling function $\Gamma(r/\xi)$.[2] We have already encountered this sort of formulation in our treatment of Koch curves [expression (1.4-1)]. This function Γ is not explicitly known, but its limit values may be determined: close to p_c, $\Gamma(x \ll 1)$ is a nonzero constant [for Eq. (3.1-21) to be satisfied] and $\Phi(x) \propto x^D$; for $r > \xi$, $\Gamma(x \gg 1)$ tends (exponentially) to zero, as there are no longer any sites in the cluster, and Φ becomes constant.

   If $g(r)$ is calculated *over the infinite cluster* when $p > p_c$, an expression of

---

[2] When L is not so large finite size effects occur. In practice these are nearly always present [see, e.g., Eq. (3.1-17)].

the form (3.1-21) is obtained, since the correlation length over this cluster is determined by the clusters of empty sites distributed within it. These have a size distribution comparable to that of the finite clusters present in the medium. The difference in this case is that for $r \gg \xi$, the mass distribution becomes homogeneous of dimension d: $M \propto r^d$, so that $\Gamma(x \gg 1) \propto x^{d-D}$.

Can any relations between the exponents be found?

Several relations (*scaling laws*) between the critical exponents have been proved or conjectured. These relations were discovered in the 1960s and 1970s when studies on scale (or dilation) invariance were being carried out, which presaged fractal structures and the renormalization group.[3] We shall introduce them as they are required.

Relations between β, ν, d and D

Consider the distribution of occupied sites in the infinite cluster within a sphere of radius ξ. On the one hand, from above, the number of sites situated in the sphere may be calculated by taking into account its fractal nature:

$$M = \xi^D \propto (p-p_c)^{-\nu D}$$

On the other, it may be observed that this number is proportional to the product of the probability of a site being in the infinite cluster, P_∞, and the number of (empty or occupied) sites present, ξ^d. Comparing these two calculations we find the scaling law

$$\boxed{D = d - \beta/\nu} \, . \tag{3.1-23}$$

This relation has been well verified numerically.

Distribution of the cluster sizes

It is also very important to know about the cluster distribution. Here again we are looking for the asymptotic behavior, that is, the distribution, n_s, of large clusters of size s. *At the percolation threshold*, because of the dilation invariance of the cluster distribution, we should expect a distribution of the form

$$n_s \propto s^{-\tau}, \quad s \to \infty. \tag{3.1-24a}$$

This behavior is also well verified numerically in those expressions where the exponent τ appears.

When $p = p_c$ no longer holds, the scaling function takes the general form

$$n_s \propto s^{-\tau} f(|p - p_c| s^\sigma), \quad s \to \infty. \tag{3.1-24b}$$

---

[3] The many authors who established the foundations of this important field are not cited here. The reader can refer to classic books.

Table III — *Percolation exponents in d = 2, d = 3 and for the Bethe lattice*
(Stauffer, 1985)

Exponent	$d = 2$	$d = 3$	Bethe	Quantity
α	$- 2/3$	$- 0.6$	$- 1$	Total number of clusters
β	$5/36$	0.42	1	Infinite cluster
γ	$43/18$	1.8	1	Mean finite cluster size
ν	$4/3$	0.88	$1/2$	Correlation length
σ	$36/91$	0.45	$1/2$	Number of clusters $p = p_c$
τ	$187/91$	2.2	$5/2$	Number of clusters $p = p_c$
$D\,(p = p_c)$	$91/48$	2.52	4	Fractal dimension
$D\,(p < p_c)^4$	1.56	2.0	4	Fractal dimension
$D\,(p > p_c)^5$	2	3	4	Fractal dimension
μ	1.3	2.0	3	Conductivity

σ may be found fairly easily. The quantity appearing in the scaling function must be dimensionless: say, of the form R_s/ξ, with ξ representing the unique scale length of the system. R_s varies as $s^{1/D}$, while $\xi \propto |p-p_c|^{-\nu}$; as a result $s^{1/D}\,|p-p_c|^\nu$ and, hence, $s^{1/\nu D}\,|p-p_c|$ are invariant. Thus,

$$\boxed{\sigma = 1/\nu D} \; . \tag{3.1-25}$$

Knowing n_s, all the mean values mentioned earlier may be calculated: P_∞, S, etc. Using the relationship, $P_\infty(p) + \Sigma_s\, sn_s(p) = p$, close to p_c, and the fact that $P_\infty(p_c) = 0$ and $\Sigma_s\, sn_s(p_c) = p_c$:

$$P_\infty(p) = p - p_c + \Sigma_s\, s\{n_s(p_c) - n_s(p)\} \; .$$

In the neighborhood of p_c, $P_\infty(p)$ is given by the singular part (for $p > p_c$) of the expression above. We make this calculation by replacing the discrete sum with an integral: after integration by parts, taking $\tau > 2$,

$$P_\infty(p) \propto \int_0^\infty (n_s(p_c) - n_s(p))\, s \; ds = (p - p_c)^{\frac{\tau-2}{\sigma}} \int_0^\infty dz\; z^{2-\tau}\, f(z) \; ,$$

that is,

$$\boxed{\tau = \beta\sigma + 2} \; . \tag{3.1-26}$$

The exponent γ of $S(p)$ may also be calculated:

---

[4] This is the fractal dimension of the very large clusters whose diameter is much greater than the correlation length of the percolation. These clusters, which are much more tenuous than the incipient percolation cluster and which have long "legs," are known as "animals." The dimension $D\,(p = p_c)$, on the other hand, corresponds to the large clusters when $p \cong p_c$ whose diameter is of the order of ξ.

[5] This is the dimension of the infinite cluster well above the threshold. It is known to be dense at distances greater than ξ (and of course fractal at distances less than ξ).

$$\boxed{\gamma = (3 - \tau)/\sigma} \ , \tag{3.1-27}$$

along with many other exponents. All these exponents relating to the geometric structure may be expressed in terms of any two of them (ν and D, for example).

Spreading dimension in percolation

This dimension (cf. Sec. 1.5.1) has been estimated by numerical simulation. The first calculations were carried out by Vannimenus *et al.* (1984). The most recent values are (see Bunde and Havlin, 1991):

$$d_f \cong 1.678 \pm 0.005, \text{ when d=2,}$$
$$d_f \cong 1.885 \pm 0.015, \text{ when d=3.}$$

The length of the shortest path in terms of the distance, L, as the crow flies is such that $L_{min} \propto L^{d_{min}}$, where

$$d_{min} \cong 1.14 \text{ in d=2 and } d_{min} \cong 1.39 \text{ in d=3.}$$

Backbone

A very important subset is the one known as the *backbone* of the percolating cluster (Fig. 3.1.8). This is the set of connected sites available for conduction (see also Sec. 4.2). Of course, there will be a backbone only at concentrations greater than the threshold.

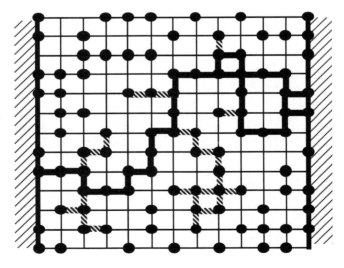

Fig. 3.1.8. The infinite cluster (here the network is of finite size) is composed of a backbone which is a simply or multiconnected set of bonds (solid line) and of dead ends (shaded lines). These dead ends contain nearly all the mass of the percolation cluster in the limiting case of an infinite network.

It also possesses a *fractal structure*, whose dimension is

$$D_B \cong 1.62 \pm 0.02 \text{ in two dimensions,}$$
$$D_B \cong 1.74 \pm 0.04 \text{ in three dimensions.}$$

Notice that it is much less dense than the percolation cluster. In particular, the probability of a site belonging to the *backbone* is given by a relationship of the form

$$P_B \propto (p - p_c)^{\beta_B} , \qquad\qquad (3.1\text{-}28)$$

with $\beta_B \cong 0.53$. On the other hand, its correlation length is equal to that of the percolation cluster itself. Lastly, its *ramification* is finite.

3.1.2 Evaporated films

Thin evaporated metallic films give rise to disordered structures of the percolating variety. Electron transmission micrographs have been taken of evaporated gold films with an amorphous Si_3N_4 substratum (Voss et al., 1982).

Fig. 3.1.9, (b) shows three connected clusters from micrograph (a) in which the concentration is p = 0.64: the set in black is the largest cluster. The authors studied layers of varying concentrations and determined the percolation threshold (which is not universal) and various exponents including τ, which gives the distribution of clusters of a given surface area. The measured value of τ is in agreement with the theoretical value in d = 2, namely $\tau \cong 2.05$.

Similarly thin layers of lead deposited on amorphous germanium have been analyzed by Kapitulnik and Deutscher (1982). The analysis of the digita-

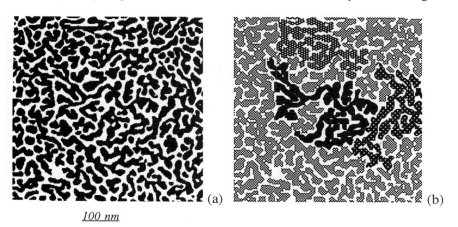

100 nm

Fig. 3.1.9. (a) Digitalized image of an evaporated gold film on an amorphous substratum of Si_3N_4. (b) From this image three clusters have been extracted. The concentration of gold coating is p = 0.64. The threshold, p_c, was found to be approximately 0.74 (Voss et al., 1982).

lized images is shown in Figs. 3.1.10 (a) and (b).

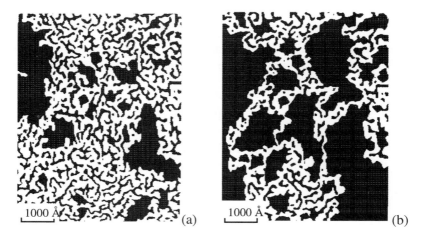

(a)

(b)

Fig. 3.1.10. Deposit of Pb on amorphous Ge. (a) Part of the infinite cluster, (b) the backbone (Kapitulnik and Deutscher, 1982).

The fractal dimension found here is D = 1.90 ± 0.02 (in the percolation model, D = 1.89...). The correlation function (i.e., the density) has also been found and this suggests a value of the exponent (D – d) in agreement with the value of D. Lastly, the backbone, the part of the infinite cluster able to conduct a continuous current, has fractal dimension $D_B \cong 1.65 \pm 0.05$, which also agrees with the percolation model.

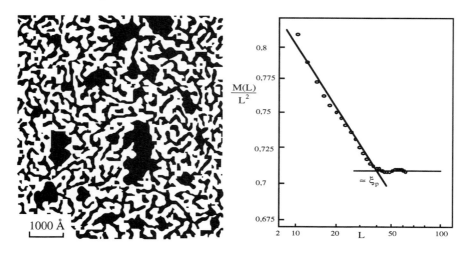

Fig. 3.1.11. Part of the infinite cluster at a covering concentration greater than the threshold. On the right is shown a log–log graph of density against size. The structure is fractal at scales up to $\xi_p \cong 40$ Å and then becomes homogeneous (constant density) (Deutscher et al., 1983)

At concentrations greater than the percolation threshold (Fig. 3.1.11), the structure is fractal at scales up to a critical distance ξ_p, while at scales greater than this distance the dimension is found to be 2. As in the previous example, the authors have calculated the distribution of clusters of surface area s, and they have found that $n_s \propto s^{-2.1 \pm 0.2}$.

3.2 Porous media

A porous medium is a randomly multiconnected medium whose channels are randomly obstructed. The quantity that measures how "holed" the medium is due to the presence of these channels is called the *porosity*, C, of the medium. Another important physical quantity, analogous to the conductivity, σ, of a network of resistors, is the *Darcy permeability*: It decreases continuously as the porosity decreases, until a *critical porosity*, C_{cr}, is rewached when it vanishes. Below this level of porosity the medium is no longer permeable (the residual porosity is said to be closed), while above it there is a path crossing the medium, i.e., an infinite "cluster."

A fractured granitic rock is an example of a poorly connected porous medium. The plane fissures can be roughly modeled by disks of random direction, position, and size.

If the distribution of channels or fissures is random and the large scale correlations in the channel structure are negligible, then transport in a porous medium resembles current flow in a random resistor network; in other words, it is related to the problem of percolation. Therefore, we might expect permeability to be characterized by a *universal behavior* that depends solely on the Euclidean dimension of the space and on the relative distance from the threshold, but which does not depend on all the microscopic details, order and nature of the connections, etc. The variation in the *permeability* close to the threshold ought to be expressible in the form

$$K = K_0 (C - C_{cr})^\mu, \quad C > C_{cr},$$
$$K = 0, \qquad\qquad C \le C_{cr},$$

(3.2-1)

where only the factor K_0 depends on these microscopic details. Of course in reality the situation is much more complex (especially due to problems of interfacial physicochemistry in the divided material). This approach via percolation theory must thus be considered as merely a powerful framework for understanding the statistics of these systems.

Furthermore, modeling a porous medium by bond or site percolation is imprecise due to the fact that the network is disordered as much from the point of view of the shape of the pores as of their interconnections (Fig. 3.2.1). However, it is known (by universality) that this does not affect the behavior (i.e., the power laws), but only the prefactors and critical values of the parameters (C_{cr} for example) which depend on the microscopic texture of the

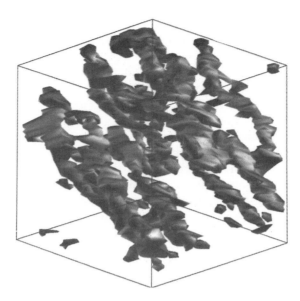

Fig. 3.2.1. An example of Fontainbleau sandstone from samples provided by C. Jacquin. The pores are light. (Image kindly provided by P.M. Adler.)

material.

Two *empirical relations* have been relatively well established for a large variety of lattices (see Table II): the product zp_{Bc} is within less than 0.06 of 2, when d = 2, and of 1.5, when d = 3, and the product f p_{Sc} is within 0.015 of 0.45, when d = 2, and of 0.15, when d = 3, where f is the filling factor of the lattice obtained by inscribing spheres of radius equal to half of the distance to the closest neighbor (f is the percentage of the total volume occupied by these spheres).

First of all, let us review some of the more interesting physical phenomena connected with porous media. These are primarily cases of transport phenomena. The question of how a fluid flows across a medium concerns monophasic flow, which is explained in Sec. 3.2.1. Another very important aspect, as much from the point of view of the fundamental physics involved as from, say, that of the extraction of petroleum from certain rocks, is the question of how a fluid injected into a medium displaces one that has already impregnated its pores; this is described in Sec. 3.2.2.

3.2.1 Monophasic flow in poorly connected media

Two main types of porosity can be distinguished. In the first type the holes are practically interconnected: this is so in the case of soils (pores not consolidated) and in the case of sandstones (pores consolidated) (Fig. 3.2.1).

The second type corresponds to nonporous, but fissured, rocks: in this case there are many connected regions of fissures ("finite clusters") with, in media that are permeable, one or more clusters extending from one end of the rock to the other (which may thus be considered as precursors of the infinite cluster of a percolating network).

The first type has already been extensively studied by physicists working on porous media. For the most part *mean field* or *effective medium* methods have been employed to homogenize the disordered medium of the pores. Various methods are possible (e.g., the self-consistent method, described above [Eq. (3.1-7)] in the calculation of $P_\infty(p)$, or a simple homogenization process), but none of them is capable of explaining the behavior of the media in the neighborhood of the percolation threshold.

Similarly, no classical method is able to describe the critical behavior of fissured rocks at the moment they become permeable. Even the self-consistent methods, which account for the percolation threshold, do not give the correct values of the critical exponents (these models imply that the permeability varies linearly with the distance from the threshold, $C - C_{cr}$, and that $\mu = 1$). It will therefore be necessary to use methods better suited to critical phenomena.

If a pressure difference ΔP is applied between two opposite faces of a porous medium with cross-sectional area A and length L, a flux, Q, of fluid is established, given by *Darcy's law*,

$$Q = \frac{K\,A}{\mu}\,\frac{\Delta P}{L} \, , \qquad (3.2\text{-}2)$$

where μ is the viscosity of the fluid. It is of the same form as Ohm's law.

The physics of monophasic flow is related to the medium's permeability, that is, to the geometry of the connected clusters (for more details consult, e.g., P. M. Adler in Avnir, 1989). If the pores are interconnected, a mean field calculation proves sufficient; however, in fissured but nonporous rocks close to critical porosity, a percolative approach is necessary. In what follows, attention will be focused on systems in which two fluids are present in a porous medium. This situation gives rise to a rich variety of phenomena in which fractal structures are often present.

3.2.2 Displacement of one fluid by another in a porous medium

When a fluid is displaced by the injection of a second fluid, the observed phenomena are naturally highly dependent on the relative properties of the two fluids.

First, they may be *miscible* (alcohol and water, etc.) or *immiscibles* (air and oil, mercury and water, etc.). Moreover, when immiscible fluids are found in porous media, one of the fluids is said to be *wetting* (i.e., it has a tendency to advance along the walls) and the other *nonwetting*: for example, in the case of water and mercury, water is the wetting fluid. The "wettability" is characterized by the angle, θ, which the interface makes with the wall (Fig. 3.2.2). A fluid is perfectly wetting when $\theta = 0$.

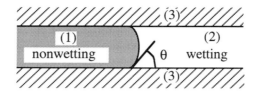

Fig. 3.2.2. Interface between two fluids (1 and 2) at equilibrium in a channel of the porous medium (3). The angle θ depends on the three surface tension couples γ_{12}, γ_{23}, and γ_{31}. The curvature of the interface induces a pressure difference between the two fluids.

An important characteristic of the interface between two fluids is the existence of an interfacial tension, γ, which induces a pressure difference, P_{cap}, on either side of the interface of these fluids meeting in a channel of radius r.

This pressure difference, called capillary pressure, may be expressed as

$$\boxed{P_{cap} = 2\,\gamma\cos\theta\,/\,r}\ .$$ (3.2-3)

Displacements of one fluid by another are governed by *Poiseuille's law* (similar in form to *Darcy's law*): the volume flux, Q_s, in a channel of length a and radius r joining two pores i and j, in which the pressures are P_i and P_j respectively, (Fig. 3.2.3) is given by

$$Q_s = \frac{\pi\,r^4}{8\,a\,\mu}\,(P_i - P_j - P_{cap})\ .$$ (3.2-4)

In this expression μ is the effective viscosity obtained by weighting the viscosities μ_1 and μ_2.

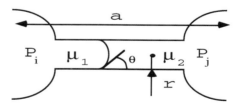

Fig. 3.2.3. Schematic representation of the channel joining two pores i and j of a porous medium in an injection experiment.

If $(P_i - P_j) > P_{cap}$, the nonwetting fluid displaces the wetting fluid ;
if $(P_i - P_j) < P_{cap}$, the opposite occurs.

Moreover, when the rate of injection is very slow the capillary forces (surface tension, etc.) dominate, whereas when it is very rapid the viscosity forces dominate.

Essentially there are three factors at play in quasistatic injection experiments (where capillary forces dominate):

(i) The external conditions imposed, such as the pressure or constant rate of flow;

(ii) the incompressibility of the fluid to be displaced (trapping problems); and

(iii) the direction of the displacement: *drainage* occurs when the wetting fluid is displaced by the nonwetting fluid (Fig. 3.2.4); *imbibition* is the reverse process.

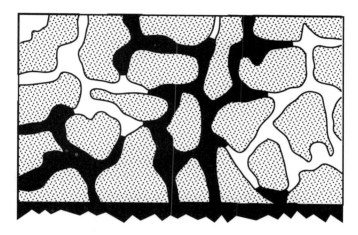

Fig. 3.2.4. Example of percolation in porous media: porosimetry by injection of mercury.

We shall not be discussing imbibition in this book since its description is slightly complicated by the presence of hysteresis phenomena (see Fig. 3.2.8), nor does it shed much further light on the aspects of greatest interest to us, the appearance of fractal structures. We mention simply that, loosely speaking, imbibition may be modeled by site percolation, in contrast to drainage which may be modeled by bond percolation, and that the imbibition process is dominated by effects arising from the presence of a thin film (wetting liquid) which acts as a precursor to the principle front (compare, for instance, Figs. 3.2.7 and 3.2.8 below).

3.2.3 Quasistatic drainage

The injection of mercury under vacuum at constant pressure is a very common method of measuring the porosity and pore sizes of a rock. A sample of rock with its interstitial water removed is placed in a vacuum and plunged into a bath of mercury: The quantity of mercury injected is measured in relation to the injection pressure.

The mercury does not wet the rock, so that it can only cross a passage between pores if the radius r is sufficiently large for the capillary pressure, P_{cap}, to equal the injection pressure, P_{inj} (Fig. 3.2.5).

At a given pressure, P, the mercury only penetrates channels whose dia-

Fig. 3.2.5 Injection of a nonwetting fluid, blocked by the presence of narrow passages of maximum dimension 2r, requiring a pressure increase in the injected fluid, P_{cap}, inversely proportional to r.

meter is greater than $(2\gamma \cos\theta)/P$ (where γ is the surface tension of the mercury and θ the angle of junction of the meniscus; $\theta \cong 40°$ for Hg); the injection of the fluid is thus, as we shall see, analogous to the search for a connected percolation cluster at the face where the fluid enters.

The corresponding percolation problem is the following: bonds for which $P_{cap} < P_{inj}$ will be considered closed (occupiable), the rest open (not occupiable). The set of clusters composed of "occupiable" bonds is constructed, but only those clusters in contact with the face through which the injection enters will be filled with fluid. Let us suppose that the network of channels is represented by a simple square lattice on each bond of which is shown the capillary radius (radius of the narrowest passage between two pores) needed to pass from one pore to the other (to simplify matters we consider a reduced radius $0 < r < 1$) (Fig. 3.2.6). Depending on the distribution of the channels' radii, there will be a greater or smaller proportion, p, of occupiable channels: for $p = p_c = 1/2$ (for the square bond lattice), the percolation threshold is attained. This proportion p_c corresponds to a critical pressure P_c, which is such that for $P = P_{inj} > P_c$ the fluid traverses all of the porous medium.

It may be observed, both experimentally and numerically, that the larger the dimensions of the rock are, the more abrupt is the transition at the threshold. Finite size effects occur since, in a section of the sample of thickness L, clusters of size greater than L allow percolation to occur and this may happen when $P < P_c$. Similarly, above P_c, impassable bond clusters (empty clusters) of size greater than L may occur, and these require P to be increased further beyond P_c for percolation to take place. Statistically, this translates to an apparent displacement of the threshold, already mentioned above [Eq. (3.1-17)].

Mercury injection at a constant rate

The rate of injection is arranged to be sufficiently slow for the pressure of the injected mercury to be the same at each point, but such that, unlike in the previous case, this pressure can vary.

We start with the system in a given state at a particular moment. As the rate of injection is constant the curvatures of all the meniscuses increase. The pressure grows until the moment when the widest passage faced by the set of these meniscuses is crossed, which momentarily fixes the pressure of all the meniscuses at the value of the capillary pressure of this passage. The fluid then invades the corresponding pore and the pressure decreases slightly; the other meniscuses in the neighboring passages recede. (This mechanism of pressure fluctuation is sometimes called a "Haines jump".) In this process of quasistatic injection, the pressures reach equilibrium and so at each moment the curvatures of the meniscuses are equal.

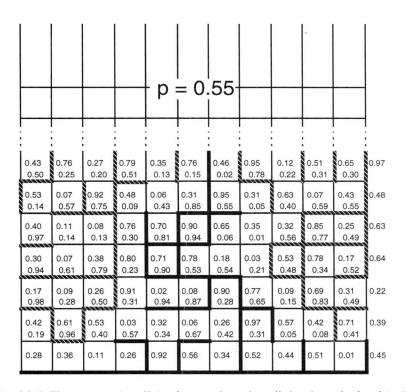

Fig. 3.2.6. The pressure is sufficiently great here that all the channels (bonds) of radius r > r₀ = 0.45 are invaded by the fluid. As the distribution of r is assumed to be uniform over the interval [0,1], a proportion p=0.55 of the bonds (in gray) are potentially active. Some of the bonds (in black), in contact with the source, have already been invaded.

Since the fluid passes through the widest passage, numerical simulation (see Fig. 3.2.6) may be undertaken by the construction of a network of passages with random connections, approximating the real medium. However, the network may equally well be chosen to be a square (d = 2) or simple cubic (d = 3) lattice because here we are concerned with universal behavior. Numbers are then allotted to the passages at random in decreasing order of radii. Each time a passage is crossed, the one with the lowest allotted number (the largest radius) is chosen next. Since the pressure is not fixed, there are no *a priori* definitively active or forbidden links. This mechanism is known as *invasion percolation*.

Remark: during a drainage process in which the displaced fluid is incompressible, the fluid may become trapped in certain regions, unlike in the situation shown in Fig. 3.2.7 where channel (2) becomes filled. This effect is negligible in three dimensions but not in two dimensions, where, if it occurs, it

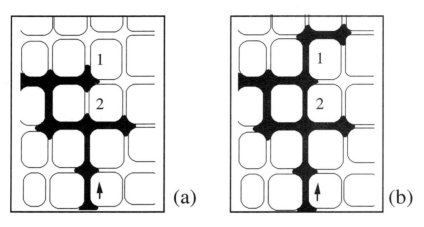

Fig. 3.2.7. *Drainage: Injection through the lower face of a nonwetting fluid (in black) displacing a wetting fluid (in white) in the channels (1), and along the walls (2). In the experimental model the channels have an average cross-sectional area in the order of a mm² and a length of 4 mm.*

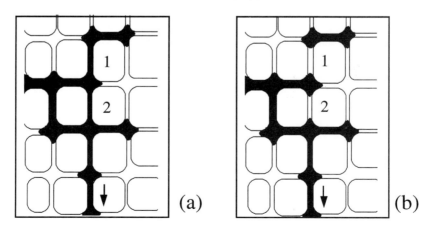

Fig. 3.2.8. *Imbibition: The nonwetting fluid (in black) is displaced here by the wetting fluid (in white) injected from above. Parts of the fluid may remain trapped (1) rendering the drainage-imbibition process irreversible; whereas channel (2) is invaded by the wetting fluid. (Cf. Lenormand et al., 1983 for further details.)*

dominates. Numerical simulations have shown that the dimension of the invaded cluster is close to 1.82, compared with a value of 1.89 for percolating clusters with no trapping (Figs 3.2.9 and 3.2.10).

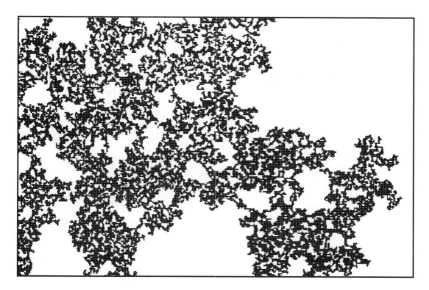

Fig. 3.2.9. Numerical simulation of invasion percolation following the method described earlier in Fig. 3.2.6. Its fractal dimension is equal to that of the infinite percolation cluster (D=91/48 in d=2) (after Feder, 1988).

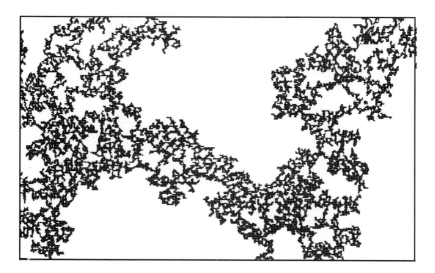

Fig. 3.2.10. Numerical simulation of invasion percolation with trapping (corresponding to an incompressible fluid some of which cannot be displaced during injection). The trapped zones dominate, and the fractal dimension is reduced to D≅1.82 (Feder, 1988).

    A precise comparison between the numerical approach and experiment has
been made by R. Lenormand and his colleagues (1988). This study concerns
drainage under various conditions, namely, quasistatic drainage leading to
capillary fingering, and rapid drainage leading either to viscous fingering or to
a stable displacement. There are two important parameters governing the
behavior of these different situations: first, the *capillary number* C_a defined by

$$C_a = \frac{Q\,\mu_1}{A\,\gamma\cos\theta}\,,\qquad\qquad (3.2\text{-}5)$$

(where fluid 1 is the injected fluid), which is the ratio between the viscous
forces and the capillary forces; and second, the ratio of the viscosities of the
fluids present,

$$M = \frac{\mu_1}{\mu_2}\,.\qquad\qquad (3.2\text{-}6)$$

Lenormand et al. (1983, 1988) have studied fluid displacement in two-
dimensional models made of transparent resin. The lattice of channels is
positioned horizontally to avoid gravitational effects. The channels have a
rectangular cross-section (1 mm x y mm), where the smaller dimension (y)
varies randomly according to a scheme similar to Figs. 3.2.7 and 3.2.8, and
they form a square lattice whose mesh size is 4 mm. On the basis of their
experimental observations, associated with the numerical simulations, the

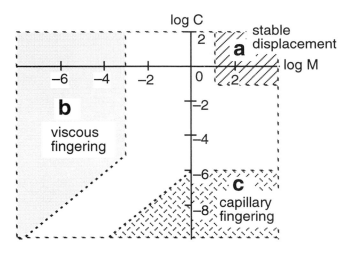

*Fig. 3.2.11. Phase diagram of the drainage of a nonwetting fluid in a porous
medium. There are three distinct bnehaviors corresponding to three distinct
universality classes. Capillary fingering, discribed in this section, is linked to the
percolation model. Viscous fingering is modeled by what is called diffusion-limited
aggregation (DLA). Lastly, stable displacement, which does not generate a fractal
structure, is modeled by anti-DLA (after Lenormand et al., 1988).*

authors have proposed a phase diagram which enables the various dynamical situations to be classified as a function of the two parameters C_a and M. This phase diagram is shown in Fig. 3.2.11. Capillary fingering, which will be explained later in this section, corresponds to low values of C_a, that is, to an injection slow enough to be considered as a series of equilibria. This is only possible for large values of M, when the viscous forces of the injected fluid dominate. This situation is modeled by *invasion percolation* described above.

Measurements of the fractal dimension found in drainage experiments, which were carried out by Lenormand and Zarcone in 1985, using these two-dimensional lattice models for various values of the capillary number C_a (Fig. 3.2.12), are in good agreement with the theoretical value D = 1.82, which is obtained when trapping is taken into consideration (except for the value, $C_a = 1.5 \times 10^{-6}$, when the method used to find D proves invalid).

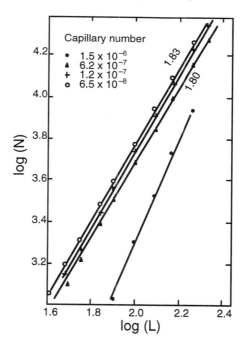

Fig. 3.2.12. Number of channels filled by air draining paraffin oil in a square lattice of size L x L, for different values of the capillary number. The slopes 1.80 and 1.83 were obtained by the method of least squares (Lenormand and Zarcone,1985).

The other two regimes (Fig. 3.2.11, regimes a and b) are not quasistatic; dynamic correlations are present between distant points in the fluid. The models are very different in this case and will be examined in Sec. 4.2, which concerns aggregation where the method of growth is closely analogous, as we shall see, to viscous fingering (stable displacement also introduces long range correlations but does not generate fractal structures).

Fig. 3.2.13 shows two plates obtained by the invasion of mercury into a two-dimensional porous model. The logarithm of the ratio M of the viscosities of mercury and of air is 1.9. The injection rate is controlled so that log C_a = – 5.9 in (a), and log C_a = – 8.1 in (b). We are well within the capillary fingering region. These structures are comparable to those of invasion percolation (Fig. 3.2.10).

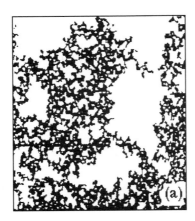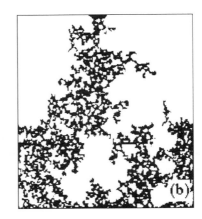

Fig. 3.2.13. Mercury (in black) displacing air (log M =1.9). The injection is slower in (b), log C_a = –8.1, than in (a), log C_a = –5.9, but the difference is not decisive. Systems (a) and (b) are both in the capillary fingering region (Lenormand et al., 1988).

3.3 Diffusion fronts and invasion fronts

When an electrical contact is created between two metals, we usually think of their interface as a regular surface. In fact it is nothing of the sort. An interface created by diffusion is in general extremely irregular. Its surface area grows indefinitely with the duration of the diffusion, as does its width; its structure eventually becomes fractal within a certain spatial domain.

To model diffusion fronts we shall take the simplest system, composed of a lattice of sites and diffusing particles which jump from site to site (Fig. 3.3.1). The latter diffuse either with a probability which is the same for all particles (noninteractive), or with one which depends on the neighboring particles (attractive or repulsive interaction). In all these cases (even the noninteractive one) the particles are taken as being approximately hard spheres such that no two particles may be present at the same site. We say that they have a "hard core."

3.3.1 Diffusion fronts of noninteracting particles

An example of this is when rare gases are adsorbed along the well cleaned face of a crystal: The crystalline surface creates a periodic potential with preferred sites where the rare gas atoms localize. It can be shown that in this lattice gas model, despite the correlation between particles due to their "hard cores," the particle density time-evolution equation is a simple diffusion equation with, in the simplest cases, a diffusion coefficient $\mathcal{D} = z\,\alpha\,a^2$, where a is the lattice spacing, z the number of neighboring sites, and α the probability of a jump per unit time.

Clearly the presence of hard cores must influence the evolution of the particles in a system which, due to the impossibility of a site being doubly occupied, cannot be the same as that of a system of independent particles. This effect appears in pair or higher order correlations, but has no influence on the properties we are considering here, which only involve the density-density correlation. This density remains uncorrelated.

The distribution of diffused particles at a given time can be determined in the following manner. First, a solution is found for the diffusion equation with the boundary conditions imposed by the problem. Then the particles are distributed over the lattice with a probability for each site equal to the calculated concentration (no correlation hypothesis). This then constitutes a realization of the actual diffusion process. In a region of fixed concentration the randomly distributed particles can then be grouped in clusters (Fig. 3.3.1) whose statistical properties are determined by percolation theory (Sec. 3.1.1). In the present case, the value of this concentration p at a given moment varies with position.

This model is known as a *gradient percolation* model: all the ingredients for percolation are present (e.g., uncorrelated site occupation), but the concen-

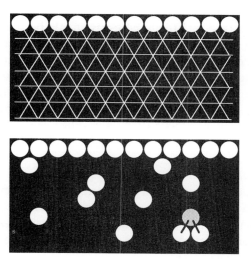

Fig. 3.3.1. Diffusion of particles over a triangular lattice. The upper figure represents the initial state: the first comb is occupied and remains so throughout the evolution. In the lower figure, some particles have diffused (the lattice has been suppressed). A cluster of three neighboring particles is shown.

tration of the sample is not uniform. A *physical property which specifies the connectivity between particles* (such as an electrical contact, or electrons jumping between atoms whose external orbits overlap, or a coupling between the nuclear spins of diffusing atoms) enables us to define connected clusters.

Taking electrical conduction as an example of such a physical property, suppose that metallic particles are made to diffuse (in direction x) in an insulating medium (see Fig. 3.3.2). Suppose that we fix the potential V_0 of the face where the diffusing metal is introduced (source). Then the set of particles in electrical contact with this source will also be at the same potential. The external surface of the collection of particles at potential V_0 is known as the *diffusion front*. Beyond this, the concentration level is too low and only clusters of particles connected to one another but not to the source are found.

On condition that the concentration does not vary too much from one extremity of the cluster to the other, these finite clusters may be assumed to be distributed like those occurring in ordinary percolation in each region of concentration p of the distribution.

Similarly, the part which is connected to the source is comparable to the "infinite cluster" of ordinary percolation or, more precisely, to a juxtaposition of "infinite" clusters, corresponding to each local concentration.

The two-dimensional case d = 2

(i) Percolation threshold

In the two dimensional case, we find that the diffusion front remains localized in a region of concentration very close to p_c, the percolation threshold associated with the lattice and connectivity under consideration.

We can understand intuitively why this should be so since the diffusion front constitutes a path running from one extremity of the sample to the other, and were the concentration to be reduced, this path would no longer exist. The threshold p_c is situated almost exactly at the center of mass x_f of the density distribution of the frontier. It has been shown that, in cases for which no exact result is known, by constructing the diffusion front, p_c in d = 2 can be determined with greater precision than any other known numerical method (see, e.g., the calculation of Ziff and Sapoval, 1986).

(ii) Front width

When a contact or an interface is created, it is important to know how its width varies as a function of the various parameters of the problem. The parameters chosen here might be the time, t, the diffusion length, $l_{\mathcal{D}} = \sqrt{(2\mathcal{D}t)}$, or perhaps the concentration gradient in the region of the front, ∇p, (which varies as the reciprocal of $l_{\mathcal{D}}$).

As the front is in a critical region ($p \cong p_c$), we should expect it to behave according to a power law in the concentration gradient. Indeed, it may be shown that the front width σ_{ft} varies proportionally to a power of $\nabla p = dp/dx$:

$$\boxed{\sigma_{ft} \propto |\nabla p|^{-\alpha_\sigma}} \qquad\qquad (3.3\text{-}1)$$

Here, $\alpha_\sigma = \nu/1+\nu$, where ν is the critical exponent of the correlation length of percolation; $\nu = 4/3$ when $d = 2$ (Sapoval et al., 1985).

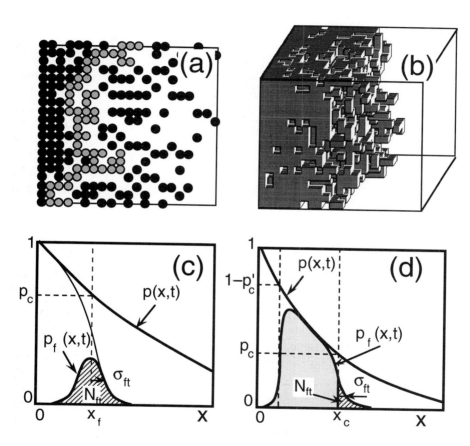

Fig. 3.3.2. *Diffusion in lattices of dimension d=2 [(a) and (c)] and d=3 [(b) and (d)]. In figures (c) and (d), p(x,t) corresponds to the concentration profile obtained under the condition $p(x=0) \equiv 1$ for all values of t. The shaded and hatched areas represent the diffusion front, the hatched part, of width σ_{ft}, being the critical region (where fluctuations are important). It is located at p_c.*

To prove relation (3.3-1) we suppose that the mean cluster radius ξ at a distance σ_{ft} from the barycentre x_f of the frontier[6] (see Fig. 3.3.2) is proportional to σ_{ft}. Close to x_f the clusters would be too large as their mean diameter ξ diverges at p_c and so then they are nearly always connected to the front,

---

[6] Or from x_c, the abscissa of the threshold p_c, since x_f tends sufficiently rapidly to x_c: $p(x_f) \to p_c$.

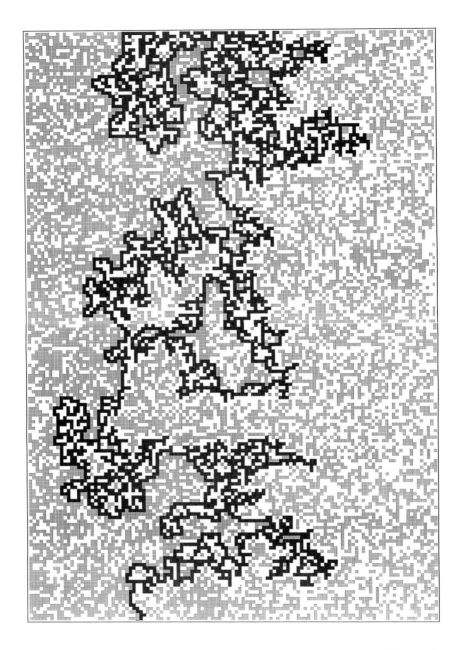

Fig. 3.3.3. Diffusion of noninteracting particles. The diffusion has taken place with
the initial and boundary conditions of Fig. 3.3.2 (a), the particles being indicated by
small squares. Only the region with concentration $0.4 \leq p \leq 0.6$ has been shown. The
chosen connection is between nearest neighboring sites: this enables us to determine
the clusters, the infinite cluster being the one connected to the source. The diffusion
front, that is, the exterior boundary of this cluster is represented in black. Its width is
σ_{ft}, its length N_{ft}, and its fractal dimension $D_f = 7/4$.

while far from x_f they become too small to be able to link up and remain finite clusters. Thus the region where some of the clusters are connected and some are not is defined by

$$\sigma_{ft} \cong \xi[p(x_f + \sigma_{ft})] \ .$$

Using Eq. (3.1-13) and developing the above relation in the neighborhood of x_f leads to Eq. (3.3-1).

(iii) Fractal dimension

In a domain restricted to the width of the front, the fractal dimension is found to be

$$\boxed{D_f = \frac{1+\nu}{\nu} = \frac{7}{4} \ \text{in} \ d = 2.} \qquad (3.3\text{-}2)$$

The value 7/4 has been confirmed both by numerical simulations and by the use of growth models or theoretical approaches involving self-avoiding walks.

(iv) The front length

The front length, N_{ft} (number of particles at the interface), also varies as a power of ∇p:

$$\boxed{N_{ft} \propto |\nabla p|^{-\alpha_N}} \qquad (3.3\text{-}3)$$

with $\quad \alpha_N = \dfrac{\nu}{1+\nu} \ (D_f - d + 1) \ .$

The proof of this result uses points (ii) and (iii) and the mass–radius relation. Consider the number of points of the front lying in a ball of radius r,

— If $r < \sigma_{ft}$ the front is fractal:

$$N_{ft}(r) \propto r^D \ .$$

— If $r > \sigma_{ft}$ then we no longer see any fractal features, the front appears like a regular Euclidean surface of dimension d–1 (a line in the present case where d = 2), and

$$N_{ft}(r) \propto A \ (r/\sigma_{ft})^{d-1} \ ,$$

where A is the number of particles, $\sigma_{ft}{}^D$, within a radius of the order of the front width.

It should be noted that although the front exhibits a change of dimension depending on whether it is observed at a large or at a small scale as do self-affine structures, it is not self-affine but self-similar, truncated by the effect of the gradient and comparable to a covering of dimension d–1 by self-similar fractal boxes of side σ_{ft}.

— The number of sites in the front for a sample of size L ($L \gg \sigma_{ft}$) is therefore

$$N_{ft} = A\,(L/\sigma_{ft})^{d-1} \approx L^{d-1}\,(\sigma_{ft})^{D_f-d+1}$$

which leads to Eq. (3.3-3).

(v) Generation of noise in $1/f^{\alpha}$

During diffusion, the front is extremely erratic and, even when the system is almost "frozen" (no observable diffusion), interface fluctuations remain important and may be a possible source of contact noise between metals or semiconductors. The behavior of this type of noise has been studied numerically in two dimensions and theoretically in two and three dimensions (Sapoval et al., 1989; Gouyet and Boughaleb, 1989). Below a cutoff frequency, f_c, the noise is white, whereas for higher frequencies the noise is in $1/f^2$. f_c also varies as a power of the concentration gradient:

$$f_c \propto |\nabla p|^{-1/(1+v)} \ . \tag{3.3-4}$$

In fact, diffusion fronts belong to the very broad class of self-organized criticality (Per Bak et al., 1988), that is to say that the structure of the front organizes itself into a critical structure on either side of the threshold p_c. It is known that these self-organized critical structures may generate noise in $1/f^{\alpha}$ as they fluctuate in time.

The three-dimensional case $d = 3$

The principle remains the same here, yet, the structure of the front is now fundamentally different. This is due to the fact that in three dimensions the presence of a percolation path no longer bars the way[7] of particles closer to the source (i.e., for $p \gg p_c$) since this path can now be bypassed, and so particles are found on the external surface at concentrations greater than p_c. Consequently, the front is now no longer localized at p_c. It extends over a range of concentrations stretching from p_c (percolation of the lattice of occupied sites) to a concentration $p_{max} = 1-p'_c$ where p'_c may be shown to be equal to the percolation threshold of the lattice of empty sites. The two lattices composed of the occupied sites and the empty sites are said to constitute a *matching pair*. For a simple cubic lattice, for example, $p_c \cong 0.312$ and $p'_c \cong 0.1$, and hence the interface extends over concentrations ranging from 0.312 to 0.9.

> The general definition of a matching pair is rather intricate. For the square site lattice, the lattices with first neighbor connectivity and with second nearest neighbor connectivity make up a "matching pair." For the triangular site lattice, the lattice with nearest neighbor connectivity is self-matching (which means that $p_c=1/2$). A matching

---

[7] Whereas in two dimensions, the particles situated on the exterior surface (or front) are all necessarily positioned on the same side of any percolating path, that is, path of touching particles joining two opposite faces of the sample.

pair of lattices in the case of a simple cube lattice are those with nearest and third nearest connectivity (see Essam, 1980 for more details).

(i) Critical region

In the neighborhood of p_c (0.312 for a simple cubic lattice), the front's structure is still critical. In particular, at concentrations weaker than p_c (the front's tail), its width is given by

$$\sigma_{ft} \propto |\nabla p|^{-\alpha_\sigma} \quad \text{with } \alpha_\sigma = \frac{v}{1+v}, \text{ where } v \cong 0.88 \qquad (3.3\text{-}5)$$

and involves a number of particles given by

$$N_{ft} \propto |\nabla p|^{-\alpha_N} \quad \text{with} \quad \alpha_N = \frac{v}{1+v}(D_f - d + 1) \qquad (3.3\text{-}6)$$

following the same power laws as in the two-dimensional case (the same proofs apply), but with a different value of v [see Fig. 3.3.2 (b)]. The critical region is fractal and has dimension: $D_f = 2.52 \pm 0.02$ (see table III, Sec. 3.1).

(ii) Ideal porous media

Over a wide range of concentrations nearly all of the particles in the infinite cluster belong to the front. This explains why the same value is found for the fractal dimensions of the front and of the infinite percolation cluster: $D = D_f \cong 2.52$. Thus, at $p \cong 0.75$ only 1 in 10^4 of the particles of the infinite cluster do not belong to the front, and this proportion decreases considerably further as p decreases (it falls to zero in the critical region). In the intermediate region the front is homogeneous with dimension $D_f = d = 3$. It could be said to constitute the *perfect disordered porous medium*, since all of its mass ($M \approx L^3$) is in its surface; of course this is quite unusual in that it is three-dimensional, but this should come as no surprise to us, as we already know of curves, such as Peano's, which have the same dimension as a surface (Sec.1.4.1), and which may therefore be thought of as a perfect two-dimensional deterministic porous medium.

3.3.2 The attractive interaction case

In the presence of an interaction between the particles, the behavior may be radically different. Indeed, below a certain critical temperature T_c a phase transition may take place. When the interaction is attractive this will be a liquid-gas transition, while in the repulsive case, which we shall not examine here, it will be an order-disorder transition. The following behavior may be noted:

At high temperatures ($T > T_c$), the structure of the front remains similar to that occurring in the noninteractive case insofar as the scaling laws concerning the width and the surface area of the front, as well as the one concerning its fractal dimension, are all preserved. Only the small scale structures change due to the correlations between particles arising from the interaction.

At low temperatures $(T < T_c)$, the stable structure of the front is no longer fractal, because an interface, whose width is fixed solely by the temperature, is established between the liquid phase and the gaseous phase. In Fig. 3.3.4 a diffused distribution at three different temperatures is shown. ßE is the relevant parameter, where ß = 1/kT, and E is the interaction energy (attractive here, so negative) between first neighbors. The transition occurs at ßE ≅ −1.76.

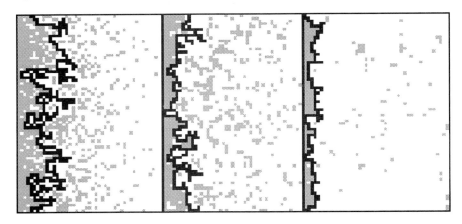

Fig. 3.3.4. Diffusion of interactive particles for three different temperatures: from left to right, respectively, βE = 0 (no interaction), βE = −1 (high temperature), and βE = −2 (low temperature). The front is only fractal for βE ≥ βE_c = −1.76... Above the critical temperature, the front has a finite width at thermodynamic equilibrium (Sapoval et al., in Avnir, 1989, p. 237).

A situation which occurs very frequently, and hence is of great interest, is when a *quench* interrupts a diffusion process. Before the quench, diffusion takes place in the high temperature region and the front is described by a *percolation gradient* model. After the quench occurs, a nucleation process develops (by spinodal decomposition). The droplets of liquid grow with time until their size approaches that of the receptacle (the phase transition has then been completed). In the regions of highest concentration these droplets form clusters, and it is then possible to verify that the model remains a percolation gradient model in which only the distance scale has been altered (renormalized): the mean droplet size, or equivalently their mean separation, replaces the intersite distance of the lattice (or the particle size for off-lattice diffusion). A structure obtained in this way by a quench starting from an equilibrium state at high temperature (the profile varies linearly from concentration 1 to concentration 0) is shown in Fig. 3.3.5 (only concentrations ranging between 0.4 and 0.6 are shown). The simulation is carried out using a Monte Carlo method and so-called Kawasaki dynamics.[8]

---

[8] In the lattice being considered, each particle is chosen at random and jumps to one of its neighboring sites, provided that it is empty, with a probability which depends on the

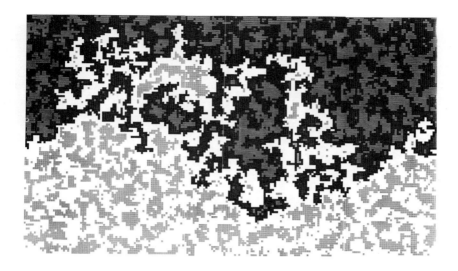

Fig. 3.3.5. Numerical simulation of attractive particles diffusing at high temperature followed by a quench. Droplets have been formed, some of which have linked up to form clusters (in gray). The interface is represented by the solid line. It has a fractal dimension of 1.75. The region situated between the source and the front has been superimposed on a gray background, (Kolb et al., 1988).

Examples of front structures observed experimentally

A remarkable analysis has been carried out on a closely related model, shown in Fig. 3.3.6 (Wool, 1988). This structure is obtained by the diffusion, in the presence of an electric field, of silver chloride molecules within a polymer layer (polyimide film), and the precipitation of silver (shown in black in the image). The nucleation of the silver is very similar to that of spinodal decomposition: the clusters grow as the reduction takes place, and their concentration is proportional to the concentration of diffused silver chloride. Wool has measured a fractal dimension $D_f = 1.74$ (against a theoretical value of 1.75) for this two-dimensional interface between the reduced silver (in black) connected to the source and the purely polymer region (in white).

The second example we shall describe here occurs when a fluid is injected into a porous medium in the presence of gravity. This example might appear somewhat remote from the case of diffusion we have just considered, but in fact this is not so. Earlier, in Sec. 3.2 on porous media, we described how drainage occurs for very slow flows. The capillary number is small and the injection may reasonably be interpreted by an invasion percolation model. Due to the presence of gravity, the fluid pressure is greater at the bottom of the region invaded by the fluid and the concentration in the pores that have been

interaction and a parameter T. This parameter is defined so that the particle distribution obtained when equilibrium is reached is equal to that of thermal equilibrium at temperature T.

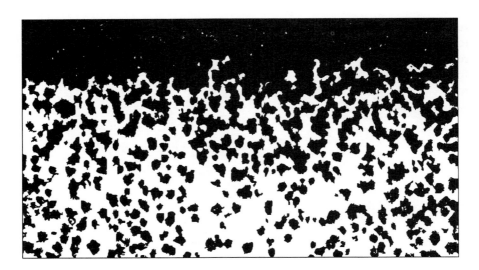

Fig. 3.3.6. Structure obtained by diffusion of silver chloride in a polymer (polyimide) and reduction of silver under the action of an electrical field across the medium. The diffusion is three-dimensional but the analysis has been carried out in a section parallel to the diffusion. The reduced silver is in black, the polyimide in white (Wool, 1988).

invaded is higher the lower their position. The appropriate model for this problem is called invasion percolation in a gradient. The underlying model is the same as in the case of diffusion in a gradient.

Invasion percolation in a gradient may be modeled as follows: in a lattice of size L^d, random numbers are selected between z/L and $1+z/L$ and assigned to each site (see Fig. 3.2.6 which occurs when $L = \infty$). In this way, the "resistance" to invasion grows linearly with z. Invasion begins at $z = 0$ and continues as in ordinary invasion percolation. The process stops before the top of the lattice is reached. The invaded region corresponds to the infinite cluster of percolation in a gradient (the concentration gradient here being $1/L$).

An experiment of this type has been carried out by Clément and Baudet in the following manner: a cylinder of diameter 10 cm and about 20 cm tall is filled with finely crushed glass (distance 200 μm between pores) as uniformly as possible. Through an orifice in the bottom a liquid metal of low melting point is injected. The metal used was Wood's metal (melting point around 70°C), which is nonwetting. The injection is carried out very slowly as sketched in Fig. 3.3.7. The injection having terminated at a certain level, the system is cooled down, the metal solidifies, and the loose grains of glass are replaced with a resin. The system may then be analyzed in detail. Sections are made at various heights, z, and the correlation function between pores is determined after digitalization. As the pressure in the Wood's metal varies with z due to gravity, so does the number of invaded pores. This can be described in fine detail using the results obtained for percolation in a gradient [the invaded region corresponds essentially to the shaded and hatched regions of Fig. 3.3.2 (d)]. The theoretical approach compares reasonably well with the experimental results (Gouyet et al. 1989). The fractal dimension of the invasion front is found to be close to 2.4 (against a theoretical value of 2.52). In principle, this method allows the distribution of pore sizes to be determined, something which few other methods manage.[9]

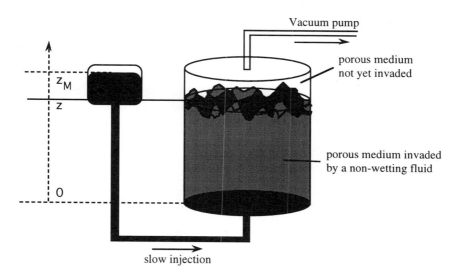

Fig. 3.3.7. Experiment consisting of the injection of a nonwetting fluid (Wood's metal) into a porous medium (crushed glass). The Wood's metal is injected extremely slowly so as to approximate capillary fingering. The gravitational field causes the pressure to vary linearly with the height of fluid injected. The theoretical model is therefore one of invasion percolation in a concentration gradient. The theoretical results obtained earlier apply remarkably well to the experimental data and allow the various parameters of the system to be found (Clément et al., 1987).

Noise generation during injection

It has long been known that injecting a nonwetting fluid into a porous medium causes fluctuations to occur, either in the pressure, if the rate of injection remains constant, or in the rate of flow, if the pressure at the injection source is slowly increased. This phenomenon is due to the fact that whenever the fluid enters a channel of radius r, the pressure is sufficient great for it to invade all those pores which are connected to this channel by channels whose radius exceeds r. Naturally, these collections of pores form the finite clusters which appear in the corresponding percolation model (see Sec. 3.2.3). In this

---

[9] Measurements of the fractal distribution of pore sizes in resins have been made using three complementary experimental techniques by Chachaty et al. (1991). These are micropores whose radius ranges between 3 nm and 1 mm.

way, the statistical noise generated by these fluctuations is very similar to the noise in $1/f^\alpha$ generated during a diffusion process (Sec. 3.3.1, v). Furthermore, injection processes also result in self-organized critical structures. In addition to the behavior of the noise, which is white below f_c [Eq. (3.3-4)] and Brownian above it, we can also calculate $N_{ev}(s)$ the number of events of size s (clusters of invaded pores) per unit time (Gouyet, 1990). This number obeys a power law which, at least for $d = 1, 2$ and $d \geq 6$, is given by

$$N_{ev}(s) \propto s^{-\mathcal{Y}_s} \quad \text{with} \quad \mathcal{Y}_s = 1 + \frac{D_f - D_{red}}{D} \qquad (3.3\text{-}7)$$

Here D_f is the fractal dimension of the invasion front (equal to 7/4, or often in practice, to 4/3), D is the dimension of the infinite cluster (equal to 91/48), and $D_{red} = 1/\nu$ is the dimension of the red links (equal to 3/4), which gives $\mathcal{Y}_s \cong$ 1.53 (or 1.31, if $D_f = 4/3$). Eq. (3.3-7) is thought to be valid in any dimensions.

3.4   Aggregates

Aggregation processes, with their very special characteristics, have been studied for a long time,[10] but it is only recently that they have started to be studied quantitatively. This is thanks to developments in two areas: fractals and the notion of scaling laws, and numerical simulations that use simple growth laws to generate aggregates. These elementary, theoretical growth processes are extremely varied and generate structures ranging from percolation to Laplacian fractals, and even the Sierpinski gasket. Consequently, for the sake of clearer exposition, we shall treat them in a separate chapter and instead dedicate this section to aggregation processes that are physically observed in nature, by presenting some experiments and the different geometries which they generate.

3.4.1 Definition of aggregation

Aggregation is an irreversible physical process in which macroscopic structures, called *aggregates*, are built up from elementary structures (particles or micro-aggregates) interacting via mutual attraction. Here we shall concern ourselves with the geometric structure of these aggregates, and later, in Chap. 4, we shall examine the important question of their growth kinetics.

In the aggregation models studied below we shall assume that the micro-aggregates or particles have already been formed, or, in other words, that the nucleation stage is over when aggregation commences. These micro-aggregates are generally fairly compact and the distribution of their sizes is relatively uniform.

---

[10] Cf., in particular, the remarkable book written by the biologist d'Arcy Thompson, *On Growth and Forms*, Cambridge, 1917 (or the reedition, 1961).

We may distinguish two major processes of aggregate growth,[11] processes which lead to structures similar to those found in other areas of physics, such as fluid injection or dielectric discharge. These two basic models are known as cluster–cluster aggregation and particle–cluster aggregation.

In the cluster–cluster process, aggregation occurs chiefly among clusters of comparable size. In the simplest model (Jullien and Kolb, 1984), two clusters of size s_0 give rise to a cluster of size $2s_0$: this is a totally hierarchical process as then two clusters of size $2s_0$ produce a cluster of size $4s_0$ and so on. In practice, systems do not remain monodisperse (i.e., only one size present at any given moment), but the nature of the polydispersity that develops is such that clusters of different sizes aggregate in a process which satisfies a global scaling law (Brown and Ball, 1985).

In the particle–cluster process, a cluster of size $s+1$ is formed from a large cluster of size s ($s \gg 1$) after the addition of a single particle. In this case, the process is highly asymmetric.

These two processes produce aggregates with very different properties. Moreover, with either process, aggregation may take place in three possible different ways: it may be limited by the reaction process, that is, a potential barrier must be overcome for sticking to take place; or limited by the diffusion process, in which case the bodies (clusters or particles) diffuse in space and stick together the moment two of them collide; and lastly, aggregation may be ballistic, the bodies move freely in space (rectilinear trajectories) before aggregating.

Table IV below displays these various combinations and the type of physical phenomena to which they apply, along with their associated fractal dimensions. Notice that reaction-limited and ballistic particle–cluster aggregation generate dense aggregates ($D = d$), only their surfaces are fractal. We shall encounter this feature again when we come to look at growth models,

Table IV: A summary of different types of growth

	Particle–cluster	Cluster–cluster
Diffusion	Electrodeposition, dielectric breakdown, injection (DLA)	Colloids and aerosols (screened)
D	1.72 (d=2) 2.50 (d=3)	1.44 (d=2) 1.75 (d=3)
Ballistic	Deposition, rough surfaces, sedimentation, etc.	Aerosols in a vacuum
D	2.00 (d=2) 3.00 (d=3)	1.55 (d=2) 1.91 (d=3)
Reaction	Tumors, epidemics, forest fires, etc. (Eden)	Colloids and aerosols (lightly screened)
D	2.00 (d=2) 3.00 (d=3)	1.59 (d=2) 2.11 (d=3)

---

[11] Which should not be confused with other deterministic growth processes such as the growth of crystals.

such as the Eden model, in Chap. 4.

The classification of the examples in table should not be taken as absolute. For example, electrodeposition may only be modeled by diffusion-limited aggregation (DLA) if the ion concentration is low (the electrochemical aspect of this phenomenon is, moreover, far from clear [12]). Injection obeys either a percolation invasion DLA model, or an anti-DLA model, depending on the values of its parameters M, the viscosity ratio, and C_a, the capillary number (see Sec. 3.2.3). Furthermore, an aggregate may be formed under the action of several different growth models: ballistic at a certain scale, diffusion at larger ones. In this case there is a change of régime in the system (a crossover) (see, e.g., Sec. 4.2.2 on the extension of the Witten and Sanders model).

In this section we are going to describe various aggregation experiments. We shall start by looking at experiments involving aggregation of aerosols and colloids, which are best modeled by diffusion or reaction-limited cluster–cluster aggregation [a detailed study of the experimental observations of aggregation carried out by M. Matsushita, as well as a paper by P. Meakin on the numerical aspects may be found in the book edited by Avnir (1989)]. We shall then go on to mention several experiments in which diffusion-limited particle–cluster aggregation is the principle ingredient. These include the examples of layers deposited by cathodic pulverization, electrolytic deposits (under certain conditions), filtration, dielectric breakdown, and the injection of one fluid into another (again only under certain conditions). It should be mentioned here that a number of experiments described by this model, which generate similar geometries, are not strictly speaking particle–cluster processes, but are nonetheless governed by the same equations (see Sec. 4.2).

3.4.2 Aerosols and colloids

Aerosols were the first aggregates to be shown to have a fractal nature; together with colloids they represent two important varieties of aggregates.

The aggregation process for aerosols is initiated by dispersing minute particles into a gaseous medium. These particles may be produced in a variety of ways: for example, smoke particles coming from the incomplete combustion of a flame; or the vaporization of a material coating a filament by rapid heating, etc.

Once dispersed throughout the gaseous medium, the particles, as long as they are electrically neutral, will behave like independent Brownian particles, except at very short distances when an attractive force (van der Waals force) causes them to group together. It should be noted however that when the gas pressure is sufficiently low (i.e., when the mean free path between collisions is large) the motion of the particles may be ballistic.

It was first observed that aggregates have fractal structures by Forrest and Witten (1979). This work was the starting point for a great many other studies which investigate the occurrence of fractal structures in aggregation and growth

---

[12] See, for example, the experimental work of Fleury et al. (1991) on this subject.

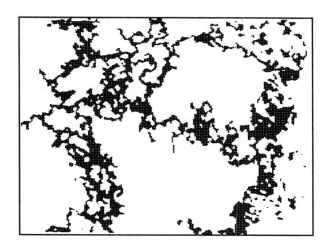

In 1979, Forrest and Witten were the first to study the distribution of iron particles produced by the vaporization of a heated filament in a cold, dense gas. The particles have a radius of 35Å. The fractal dimension D=1.51±0.05 is obtained from the mean of the masses in squares of increasing size centered randomly in the aggregate.

Fig. 3.4.1. Image of iron particle aerosols produced by a transmission electron microscope (Forrest and Witten, 1979).

phenomena. Inspired by the theory of critical phenomena, Forrest and Witten were the first to interpret aggregation in terms of scaling laws through their experience in dealing with aggregation of iron particles. They worked out the density–density correlation function g(r), the mean centered value of the product of the densities, averaged over all pairs of points, a distance r apart, (the aggregation being assumed to be isotropic and homogeneous),

$$g(r) = \langle \rho(\vec{r}_0 + \vec{r}) \, \rho(\vec{r}_0) \rangle - \langle \rho(\vec{r}_0) \rangle^2, \qquad r = |\vec{r}| \qquad (3.4\text{-}1)$$

This correlation function decreases according to a power law in r,

$$g(r) \propto r^{-A} \quad \text{where } A \cong 0.3,$$

and this relation holds for various materials, suggesting a universal exponent [the measurements are made in $d = 2$ from the digitalized image of the aggregate (Fig. 3.4.1)]. Clearly, the mass of aggregate present within a radius R in a medium of Euclidean dimension d is given by

$$M(R) \propto \int_0^R g(r) \, r^{d-1} \, dr \propto R^D . \qquad (3.4\text{-}2)$$

This relates A to the fractal dimension D: $D + A = d$. The fractal dimension obtained by Forrest and Witten from g(r) was 1.7 ± 0.02 and from the mass–radius relation 1.51 ± 0.05 (the discrepancy comes from the small cluster size and the difficulty to perform good averages in the mass-radius relation).

Iron particles aggregate via a diffusion-limited cluster–cluster process (in air, iron particles move along Brownian trajectories), or, if the mean free path is not small relative to the size of the aggregates already formed, via a ballistic cluster–cluster process. The values obtained for the fractal dimension suggest that a diffusion-limited process should be chosen here.

Colloids are composed of spherical microparticles (of metal, silica, etc.), a few nanometers in diameter, in suspension in a liquid. The surfaces of these particles are electrically charged and so they remain in suspension under the action of their mutual repulsion. In the absence of charged particles dissolved in the liquid, the van der Waals attraction is masked by their Coulomb repulsion (due to the electrical charges carried by these particles). In this way, the superposition of the van der Waals attraction and the Coulomb repulsion sets up a potential barrier ε (Fig. 3.4.2) which is generally greater than kT and hence impassable. Such metastable solutions can survive for a very long time; indeed, there are colloids made by Faraday more than 130 years ago still in

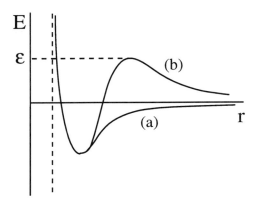

Fig. 3.4.2. Potential between two charged particles distance r apart. When the ions are not screened, (b), they repel each other and the barrier is sometimes sufficient for a colloidal solution to survive tens of years or even a century. By adding a salt to the solution the screening changes, (a), and the colloid is then no longer metastable. Aggregation will then occur.

existence today. The addition of a salt to the liquid introduces free ions oppositely charged to the particles in suspension, and partially or totally screens the repulsive interaction. When ε is of the order of, or is less than, kT, aggregation begins and the solution becomes unstable. The aggregates increase in size and start to diffuse light by the time they measure about a micron across. This effect is signalled by the appearance of an opalescence.

The cluster size distribution is very sensitive to the height of the potential barrier. Aggregates are formed in the following manner: so long as the colloidal particles are distant from each other, they diffuse in a Brownian way, since the screening suppresses their Coulomb repulsion (i.e., at large distances they do not interact).

When the screening is total, two particles approaching each other at a distance close enough for a strong van der Waals attraction to occur between them, will irreversibly stick to one another. The doublet thus formed continues to diffuse in the liquid where it meets other particles or doublets which then combine to create a larger cluster, and so on. Aggregation occurs rapidly in this situation, each encounter between clusters causing them to stick to each other first time.

If the screening is not total and the potential barrier, ε, is a little higher than kT, many attempts must be made before sticking takes place (the probability of sticking is given by Arrhénius' law). These unsuccessful attempts profoundly modify the structure of the aggregates which are in this case much more compact than when the screening is total. The fractal dimension and other geometric and kinetic properties are thus directly related to the mechanics of the aggregation process. In the first case, aggregation is diffusion-limited, while in the second it is the reaction (crossing the potential barrier) which limits aggregation (see Table IV above). In the latter case, aggregation is therefore much slower.

Weitz's experiments on aggregation in gold colloids are among the most accurate determinations of the scaling properties of irreversible growth processes. To carry out these determinations he used transmission electron micrographs. Gold colloids are especially well suited to the use of these techniques, since they produce large, widely contrasting aggregates. The gold particles are obtained by reducing $Na(AuCl_4)$. The sol thus produced is stable and contains particles around 14.5 nm in diameter, which are very uniform in size. Aggregation begins when pyridine is added, which has the effect of displacing the surface charges of the particles. The aggregates obtained at the end of this process are about a micron across.

The micrograph shown in Fig. 3.4.3(a) indicates that this is a diffusion-limited cluster–cluster process; the aggregates are all about the same size. In an aqueous solution conditions for aggregation are such that the aggregates form in less than a minute. This shows that they are sticking to each other at the first encounter. The mass varies as a function of the radius as follows:

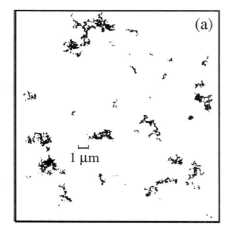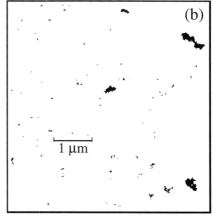

Fig 3.4.3. Photographs of gold colloids taken by Weitz using an electron microscope where (a) aggregation is fast (strong screening) and (b) aggregation is slow (weak screening) (Weitz and Lin 1986; see Matsushita in Avnir, 1989).

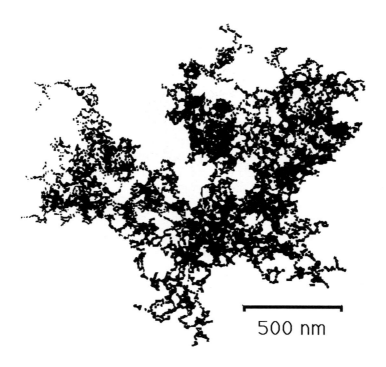

500 nm

Fig. 3.4.4. Image from a transmission electron microscope of a gold colloid aggregate produced under conditions allowing fast, or, diffusion-limited, aggregation (Weitz and Oliveria, 1984).

$$M \propto R^D, \quad \text{where } D \cong 1.78,$$

the sign of a fractal structure. The aggregate is self-similar and so has no scale lengths other than the particle and cluster sizes at any given moment. Figure 3.4.4 is an image of one of these clusters taken by a transmission electron microscope.

The images shown in Figs. 3.4.3(b) and 3.4.5 are from an experiment carried out by Weitz and Lin, in which the addition of less pyridine to the sol causes the potential barrier to be only *partially screened*. We can see that the aggregates here are more compact. This variety of cluster–cluster aggregation is reaction-limited (RLCA — reaction-limited cluster aggregation). The dimension is found to be $D \cong 2.05$ (Weitz et al., 1985). Note that, as this dimension is greater than 2, it cannot be obtained numerically from the two-dimensional image generated by projecting the cluster onto a plane, because such an image would be densely filled by particles (in Fig. 3.4.5, the holes in the central region of the cluster are of relatively uniform size, indicating a non-fractal projection, hence a dimension $D \geq 2$).

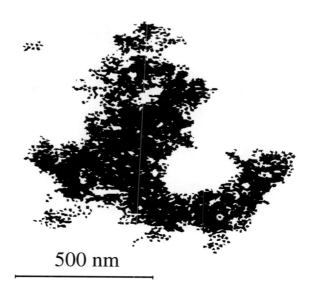

500 nm

Fig. 3.4.5. Image from a transmission electron microscope of a gold colloid aggregate produced under conditions of slow, or, reaction-limited, aggregation (Weitz et al., 1985).

Fortunately, there are other methods of displaying the fractal nature of colloidal aggregates when they are of dimension $D > 2$. One of these, employed experimentally by Clément and Baudet (Fig. 3.3.7), consists in examining plane sections (whose dimension is $D - 1$). Of course, this method would not work here due to the aggregates' fragility and their small size. Other methods employ techniques such as nuclear magnetic relaxation (Chachaty et al., 1991), but one of the most recent techniques relies on small angle scattering.

The use of scattering at small angles to determine fractal dimensions

Experiments involving the scattering of light, X rays, or neutrons are very well suited to investigate the fractal geometric characteristics of those samples whose two-dimensional images cannot be digitalized. Currently they are being used to study aggregates as well as samples of gels (especially silica aerogels).

The scattering intensity depends principally on variations in the density. Thus, for materials displaying a fractal structure, the intensity will depend on the nature of that structure. Pfeifer has proposed that three broad classes of structure should be distinguished: mass fractals, pore fractals, and surface fractals (Fig. 3.4.6). As regards wave scattering, the difference between mass fractals and pore fractals is relatively slight. This is because a mass fractal becomes a pore fractal, and vice versa, simply by interchanging the regions of matter and empty space.

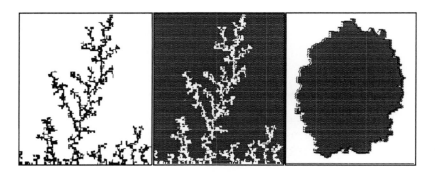

Fig.3.4.6. From left to right a mass fractal, a pore fractal, and a surface fractal. The matter is shown in black or gray, the empty regions in white. The surface fractal (an Eden cluster) is dense in its interior (D = d). A surface fractal very often has a self-affine structure.

The scattering experiment is set up as shown in Fig. 3.4.7. The use of light, X rays and neutrons allows the sample to be explored at wavelengths ranging from 1 Å to 1 μm. When experiments are carried out on aggregates they are generally found to be mass fractals.

The intensity of radiation scattered at small angles (Q small) by a mass fractal as a function of the fluctuation Q of the wave vector is given by

$$I \propto Q^{-D} \quad , \tag{3.4-3}$$

where $Q = 4\pi \lambda^{-1} \sin(\theta/2)$. The fluctuations in the wave vector are related to the fluctuations in the local density; a lack of uniformity in the sample causes the wave to scatter.[13] Thus, the density variation in a fractal structure yields

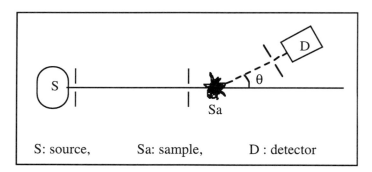

Fig. 3.4.7. Schematic diagram of the measurement of small angle scattering. The source emits radiation of wavelength λ (visible light, X rays, or neutrons). Experimentally θ is varied, while λ is held constant.

---

[13] The tiny isolated (approximately spherical) particles give a law in Q^{-4}, for values of Q of the order of the reciprocal of the particle size; this is Porod's law.

Eq. (3.4-3). This relation is derived by taking the Fourier transform of the density–density correlation function g(r). More precisely,

$$I(Q) = A \int_0^\infty g(r) \frac{\sin Qr}{Qr} r^2 \, dr \qquad (3.4\text{-}4)$$

where Q is the wave vector length.

The relation in Q^{-D} is valid as long as 1/Q remains within a domain ranging from the size of a particle to the cluster radius (to be exact its radius of gyration or correlation length). It tells us, therefore, not only about the fractal dimension of the scattering object, but also about the extent of the fractal domain.

This small angle scattering technique was the one used by Weitz and Oliveria to determine the fractal dimensions of the clusters shown in Figs. 3.4.4 and 3.4.5. They found dimension values, D = 1.77 ± 0.05 for diffusion-limited aggregation, and D = 2.05 ± 0.05 for reaction-limited aggregation. Other experimenters have obtained similar results in agreement with the results produced by numerical simulations (which give 1.75 and 2.11, respectively; see table IV in Sec. 3.4.1).

This technique has also been used on silica aggregates (see, e.g., P.W. Schmidt in Avnir, 1989). Silica and alumino-silicate aerogels are the most suitable materials for this type of work (Vacher et al. 1988; Chaput et al., 1990). Aerogels have a mean density ranging from 5 kg/m^3 to 250 kg/m^3, while the particles that make them up have a density close to 2100 kg/m^3. Since the mean density of a fractal tends to zero as its size increases, the lighter the aerogels are, the larger is the range of their fractal region. Furthermore, aerosols also have remarkable thermal insulating properties. We shall meet them again in Chap. 5 when we study transport and vibration modes in fractal structures.

Fig. 3.4.8 shows the variation of I(Q) for several different aerogels densities, and demonstrates the progressive appearance of increasingly extended fractal structures as the mean density decreases. The curves display two changes of régime: one when $Q \cong 0.07 \text{Å}^{-1}$, corresponding to scattering off the constituent particles of the material, producing, for large Q, a Porod régime; the other for values of Q between 10^{-2} and 10^{-3}, corresponding to the correlation length of the fractal régime (and so directly related to the density).

In general, for a fractal whose volume is of dimension D and which is bounded by a surface of dimension D_s, it can be shown that the intensity takes the form

$$\boxed{I \propto Q^{-2D + D_s} \qquad \text{with} \quad d - 1 < D_s \le D < d.} \qquad (3.4\text{-}5)$$

From this expression we can recover the result for scattering off a mass fractal, as this corresponds to $D = D_s$, as well as for scattering off a nonfractal object for which D = 3 and $D_s = 2$ (Porod's law).

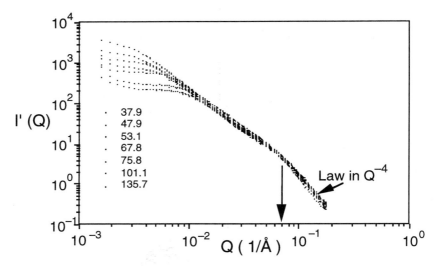

Fig. 3.4.8. Graphs of small angle neutron scattering exhibiting the self-similarity of samples of varying density. The densities, ρ, are shown in the figure in kg/m³ in the same order as the graphs. The intensity has been normalized: I'(Q) = I(Q)/ρ. This reveals the remarkable way in which the graphs overlap within the fractal region. The left-hand limit of the fractal region occurs where the slope changes abruptly (from 100 to 400 Å). The arrow corresponds to the size of the constituent particles, around 14 Å (Chaput et al., 1990).

For surface fractals of dimension 3, we find

$$I \propto Q^{-6+D_s}, \quad \text{with } 2 < D_s < 3. \tag{3.4-6}$$

Surface fractals are found in a number of different growth processes, both ballistic and reaction-limited (cf. Table IV in Sec. 3.4.1, and Chap. 4 on growth models). In particular, scattering experiments have been carried out on soots produced by aggregation of silica and on colloids. In this way, the fractal dimensions of their rough surfaces can, in principle, be determined (see Schaefer, 1988). Interpretation of these results is rather difficult due to the presence of large regions where the régime is changing (i.e., passing from a surface fractal regime to a mass fractal or Porod régime).

We should also mention that neutron scattering techniques have displayed fractal structures occurring in natural rocks. Experiments on many sandstones and clay schists have shown their behavior to be governed by Eq. (3.4-6), and their dimension D_s up to 500 Å to be between 2 and 3 (cf. the article by Po-zen Wong, 1988).

3.4.3 Macroscopic aggregation

There are other aggregation processes which are more readily observable. For instance, when particles are held in suspension in a liquid, they may reach up to a millimeter in size (unlike colloid and aerosol particles which may be much smaller than a micron). When the particle interactions are negligible (in particular, in relation to the forces generated by the liquid's motion), aggregation does not take place and the main problem, which is extremely important for studying lubricants, is to try to understand the various characteristics (viscosity, etc.) of the heterogenous fluid. When the particle interactions are too large to be ignored, aggregation also occurs, on top of which are superimposed hydrodynamic effects due to the motion of the carrier liquid (shearing, etc.).

Here we mention a couple of experiments involving two-dimensional macroscopic aggregation, one performed at the Ecole de Physique et Chimie in Paris, the other at Marseille. In the experiment carried out by Allain and Jouhier (1983), balls of wax of diameter 1.8 mm were placed on the surface of a tank of water. Due to the surface depression caused by each ball, there is a short range (of a few millimeters) attraction between them (the mattress effect). The experiment commences by agitating the water to disperse the balls over its surface. As soon as the agitation dies down, aggregation can be observed and two-dimensional structures of dimension $D \cong 1.7$ are produced, compatible with a cluster–cluster DLA model (DLCA). In the experiment carried out by Camoin and Blanc (Camoin, 1985), aggregation was observed under shearing: when there is no interaction between the microspheres,[14] the clusters generated have the structure of a "stirred" percolation cluster, producing nearly mean-field behavior; in the presence of an attractive interaction, there is a critical shearing rate above which the hydrodynamic effects start to dominate the capillary effects. Thus, in the presence of strong shearing, the clusters obtained are compact.

All the processes that we have just described are cluster aggregation processes. To observe clusters growing via particle–cluster aggregation, we must have (fixed) nucleation centers distributed sufficiently far apart so that the forming clusters do not influence each other. Even here, growth may be diffusion-limited, reaction-limited or ballistic. There now follow several examples of aggregation which are more or less adequately modeled by either a diffusion-limited particle–cluster process, a process equivalent to DLA (see diffusion-limited agregation in Chap. 4) or a ballistic particle–cluster process, as in the case of sedimentation.

3.4.4 Layers deposited by sputtering

Under certain conditions, DLA-type structures may be observed in layers deposited by sputtering. The photograph in Fig. 3.4.9 shows such a structure. Elam and his collaborators (1985) produced it by depositing a thin layer of $NbGe_2$ onto a substrate of quartz. The $NbGe_2$ atoms adsorbed on the

---

[14] To modify the interaction between spheres a thin layer of viscous liquid (vaseline oil for example) is added to the carrier liquid (which is pure or salt water). When the thickness of this layer is of the order of the diameter of the particles, the interaction between them becomes negligible.

Fig. 3.4.9. Example of a structure obtained by depositing NbGe$_2$ onto a substrate of quartz (from Elam et al., 1985).

quartz diffuse until the moment they start sticking to one of the clusters growing from a nucleation center. A cluster may be thought of here as possessing a double structure: it is made up of a skeleton obtained by reducing each branch to a simple line, which is then increasingly thickened towards the tips. In this way, the authors found a fractal dimension of about 1.88 for the NbGe$_2$ structures, but a dimension $D \cong 1.7$ for the skeleton, a value which agrees with the particle–cluster aggregation model. The clusters do not aggregate among themselves, but even so, the nucleation centers are possibly not far enough apart for all mutual influence to be ignored during the final stage of growth. We are a long way from having a clear picture of this type of growth phenomenon.

3.4.5 Aggregation in a weak field

When an external systematic force (i.e., one other than a particle interaction) is applied to the particles, there is said to be aggregation in a field. The aggregates obtained, however, vary greatly depending on the type of force acting on the system. It may be a truly external force (that is, not involved in the growth process) such as gravity or the driving force due to the carrier fluid. The following paragraphs provide some examples.

Sedimentation

The particles form aggregates under the action of gravity. Their motion is thus essentially ballistic. The less the particles diffuse on the surface of the aggregate, the looser the structures formed. We shall describe them in greater detail in Sec. 4.3 on rough surfaces.

Filtration

The particles carried by a moving fluid congregate around the holes of a filter (as when dust aggregates in the openings of a ventilator). Here again the ballistic aspect dominates.

But in other situations such as electrodeposition, dielectric breakdown, or the rapid injection of one fluid into another more viscous one—phenomena which have many characteristics in common—the (electric or hydrodynamic) force may vary with the growth of the cluster itself. We shall see in Sec. 4.2 that the model common to this method of growth is, to a first approximation, diffusion–limited aggregation (DLA). The structures obtained are also fairly similar to those produced by Elam et al. (Fig. 3.4.9) where no external field is present.

Electrolytic deposits

These are obtained by depositing metallic atoms onto the cathode during electrolysis of a metallic salt (Fig. 3.4.10). These extremely fragile structures are even more ramified when a dilute solution and a weak current are used.

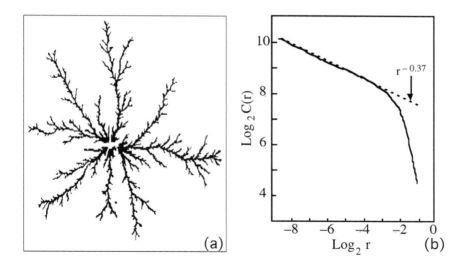

Fig. 3.4.10. (a) Electrolytic growth of metallic zinc in a thin layer of electrolyte and (b) its correlation function giving a dimension D = 2 – 0.37 (Matsushita et al., 1984).

Dielectric breakdown

Similar structures are observed when an electric field is applied between a central electrode and a circular electrode separated by an insulator (with the same geometry as in the electric deposition experiment) (Niemeyer et al., 1984). Strictly speaking, there are no particles being displaced but only molecules being polarized until the breakdown threshold is reached (Fig. 3.4.11).

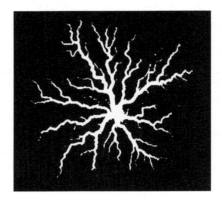

Fig. 3.4.11. Time-integrated photograph of a surface leader discharge on a 2 mm glass plate, placed in gaseous SF_6 at a pressure of 3 atmospheres (Niemeyer et al., 1984).

Fluid injection

In around 1984, Paterson noticed that when a fluid of low viscosity is injected into a radial Hele–Shaw cell filled with a viscous fluid, structures may be observed whose growth is governed by the DLA model (Paterson, 1984). These structures appear, for instance, when low viscosity fluids are injected so as to displace a very viscous fluid inside a porous medium, as shown in Fig. 3.4.12. This corresponds to the viscous fingering in the phase diagram (Fig. 3.2.11) proposed by Lenormand et al. (1988), which wespoke of earlier. The

Fig. 3.4.12. Water displacing a polysaccharide solution (scleroglucane) in a radial cell. The viscosity ratio is $M \cong 1/1500$; the rate of flow 20 ml/mn. The Hausdorff dimension measured is $D_f \cong 1.7 \pm 0.05$ (Daccord et al., 1986).

porous nature of the medium introduces a random aspect which is super-imposed on the hydrodynamic process. The more prominent the local randomness, the greater the ramification of the fractal structure.[15] In the images shown in Fig. 3.4.13, the boundary conditions are no longer radial, as was the case in the previous figures. However, the geometric characteristics, such as the fractal dimension and the multifractal structure, are similar.

Daccord and Lenormand (1987) have obtained beautiful structures clearly belonging to the same universality class, that is, diffusion-limited aggregation, by injecting a reactive fluid into a soluble porous medium. This involves the injection of a fluid into a porous medium together with a chemical reaction. In fact, the authors injected water into plaster. The final structure, corresponding to the region of plaster dissolved, is reproduced by injecting Wood's metal (melting point 70°C). One of these three-dimensional structures is shown in the color plates at the end of this book.

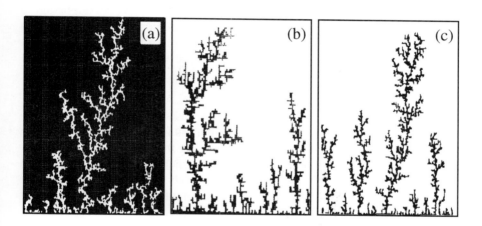

Fig. 3.4.13. Comparison of the structures obtained by (a) injecting a nonwetting fluid in the viscous fingering regime into a porous medium; (b) numerical simulation from the fluid's transport equations in the canals of a porous medium; (c) numerical simulation of diffusion-limited aggregation (Lenormand, 1989).

Let us bring these examples to a close by mentioning the beautiful experiments of Fujikawa and Matsushita (1989) where bacteria colonies, bacillus subtilis, are grown on agar plates in a concentration field of nutritional matter. This involves a growth process limited by the diffusion of the nutritional product within the substrate, and generates structures very similar to those that we have just described (the fractal dimension is of the order of 1.73 ± 0.02).

In this section we have introduced a variety of new models (Eden, DLA,

---

[15] The same experiment carried out in a Hele–Shaw cell made from two smooth plates of glass leads to structures with well rounded fingers.

etc.) without describing them in detail. This purely descriptive approach does not allow us to understand how the DLA model relates these diverse phenomena: diffusive aggregation, dielectric breakdown, viscous instability. Chap. 4, which studies growth models, will partially fill this gap. When Witten and Sander put forward the DLA model, which we shall study in that chapter, they intended it to model experiments on iron aerosols carried out by Witten and Forrest. In fact diffusion-limited aggregation does not apply to this situation, nor does it apply, generally speaking, to experiments on colloids or aerosols. Readers should find the film on aggregation processes made by Max Kolb (1986) of great interest.

3.5 Polymers and membranes

Polymers and membranes are further examples of structures involving fractal geometries. Here again, the presence of fractal structures can be put down to the large number of configurations that these objects can adopt and to the competition between energy and entropy, which most frequently causes a transition to occur. Linear polymers in solution form a fractal structure ($D \cong$ 1.67) at high temperature and a dense structure ($D = 3$) below a certain temperature known as the theta point.

3.5.1 Fractal properties of polymers

Polymers are formed by monomers linking up together in the course of a chemical reaction. This chemical reaction may take place at any of a number of sites on each monomer; the number of sites permitting links is called the *functionality* of the monomer. This process generates ramified structures with irreversible links. If the functionality of all the monomers is 2, the polymer formed is a chain. In other cases (f >2) the polymer is said to be *branched*.

As with percolation, when the lattice of polymerized molecules is sufficiently extended so as to form a very large cluster (*infinite cluster*), we describe it as a *gel*. Gels have an elastic structure (think of the pectine gels used in certain food products), while a lower level of polymerization results in a more or less viscous liquid known as a *sol*.

Linear polymers

These are produced from bifunctional monomers such as

polyethylene: ...— CH_2— CH_2— CH_2...

or polystyrene: CH_2—CH— CH_2—CH— CH_2—CH—

Chains are formed when monomers, initially dissolved in a solvent, start to connect. These chains are flexible because they can rotate about their links, and can be very long. A random walk may be taken as an initial idealized model of a chain, in which each step corresponds to a link and any two successive links can point in arbitrary directions. This type of random walk was described in Sec. 2.2.1. Its fractal dimension is 2 whatever the value of d. When we study the structure of a polymer formed in this way, i.e., from chains dissolved in a solvent, we find the small angle scattering intensity (Sec. 3.4.2, p. 131) to be of the form $I \propto Q^{-2}$.

Self-avoiding random walks

There are, in fact, constraints present for each individual chain and between chains. The interaction $V(r)$ between two (electrically neutral) monomers has a similar form[16] to the one that exists between two totally screened colloidal particles, represented in Fig. 3.4.2(a). Without entering into details, let us simply say that a useful parameter for describing the effect of this interaction in a solvent at temperature T is (cf. de Gennes, 1979, p. 56),

$$v = \int \{1 - \exp(-V(r)/kT)\} \, d^3r. \qquad (3.5\text{-}1)$$

A *good solvent* ($v > 0$) is one in which the "hard core" monomer interactions are dominant: a polymer in a good solvent should in this case be modeled not by a standard random walk, but by a *self-avoiding random walk*, that is, a random walk that never intersects itself.

The theoretical literature on self-avoiding walks is very large (cf. de Gennes, 1979 and references therein). Here we shall merely describe the very illuminating mean field approach due to Flory (1971). For a linear polymer of mean radius R in a good solvent, the free energy F contains a repulsive term F_{rep} due to the "hard core" monomer interactions and an entropic term due to the multiplicity of possible configurations. If C is the local density of the monomers, the repulsive energy is locally proportional to C^2, so that

$$F_{rep} \cong kT \ v \int C^2 \ d^dr = kT \ v \ \langle C^2 \rangle \ R^d \cong kT \ v \ \langle C \rangle^2 \ R^d,$$

where (using a mean field approximation) the mean of the square has been replaced by the square of the mean, and d is the dimension of the space. The mean density $\langle C \rangle$ is given by the ratio N/R^d, of the (fixed) total number of monomers to the volume occupied by the polymer. Hence,

$$\frac{F_{rep}}{kT} \cong v \ \frac{N^2}{R^d}. \qquad (3.5\text{-}2)$$

On the other hand, the entropic part of the free energy $F_{ent} = -TS$ (which produces a restoring elastic force) is calculated by determining the number of possible configurations of a random walk of size R (see Sec. 2.2.1). The size distribution has the form

---

[16] The interaction between monomers a distance R apart is generally of the Lennard–Jones (12-6) type, that is, it is the difference between a term in $1/R^{12}$ and a term in $1/R^6$.

$$p(R,N) = \frac{A}{N^{d/2}} \exp -B \frac{R^2}{Na^2}$$

and the free entropic energy (proportional to the logarithm of the probability)

$$\frac{F_{ent}(R)}{kT} \cong \frac{F_{ent}(0)}{kT} + \left(\frac{R}{R_0}\right)^2 . \tag{3.5-3}$$

R_0 is the ideal mean radius (in the absence of an interaction), hence $R_0^2 \propto N$. The size of a "real" polymer in equilibrium is found by minimizing the total free energy

$$\frac{F - F_0}{kT} \cong \left(\frac{R}{R_0}\right)^2 + \frac{v\,N^2}{R^d} . \tag{3.5-4}$$

Minimization with respect to R gives

$$\boxed{N \propto R^D \quad \text{with } D = \frac{d+2}{3}} . \tag{3.5-5}$$

The structure of *a polymer chain in a good solvent is fractal*, its *mean field* fractal dimension is D = 4/3 in d = 2 and D = 5/3 in d = 3. These values are remarkably accurate (less than 1% from the best numerical values). In two dimensions, the "exact" approach using the conformal invariance gives D = 4/3.

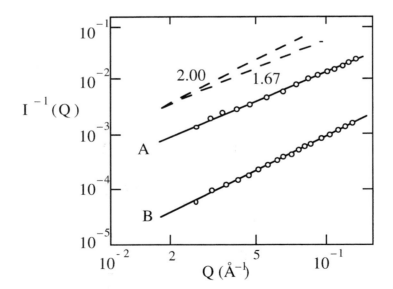

Fig. 3.5.1. Experimental data of small angle neutron scattering off polystyrene chains dissolved in benzene (curve A): the gradient 5/3 is shown as a dashed line. In the case of molten polystyrene (curve B), the gradient attained is 2, since screening causes a chain in a molten polymer to be modeled by a random walk (Farnoux's thesis, cf. Daoud and Martin in Avnir, 1989).

Experiments with neutron scattering have been carried out on polystyrene dissolved in benzene. This solvent may be considered as a good solvent, and the scattering intensity is found to follow very closely the relationship $I(Q) \propto Q^{-D}$ with $D = 5/3$ (Fig. 3.5.1).

This is so as long as the polymer is dilute enough in the solvent for the interaction between chains to be neglected. If the monomer concentration C is steadily increased, a concentration C* is reached at which the polymers overlap. This concentration is equal to the concentration of monomers in a chain, that is,

$$\boxed{C^* \cong \frac{N}{R^d} \propto N^{1-d/D} \propto N^{-4/5} \quad (\text{in } d=3)} . \tag{3.5-6}$$

Scaling laws

As C* plays the role of a critical concentration, we should expect the dimensionless ratio C/C* to be the relevant parameter in problems involving polymers dissolved in good solvents. Any property dependent on N and C may thus be written

$$F(N,C) = F(N,0) \, f(C/C^*)$$

or alternatively,

$$F(N,C) = F(N,0) \, f(CN^{4/5}) . \tag{3.5-7}$$

Polymer chains in a bad solvent

This is how we describe the case where $v < 0$, for then the free energy defined by Eq. (3.5-4) no longer possesses a minimum. The series must be expanded to higher order, since the term for the monomer interaction (in C^2) is now negative, that is, dominated by the attractive part of the interaction. The term in C^3 (for three bodies) remains positive and v vanishes at the transition temperature $T = \Theta$:

$$\frac{F - F_0}{kT} \cong \frac{R^2}{R_0^2} + \frac{v \, N^2}{R^d} + \frac{w \, N^3}{R^{2d}} + \cdots . \tag{3.5-8}$$

For temperatures below the Θ point ($v < 0$), the polymer "collapses" to a dense structure and $D = d$. All orders (n-body interactions) now play a role.

Polymer chains without a solvent and vulcanization

A plate of spaghetti provides a good picture of a collection of chains at concentration $C = 1$, i.e., with no solvent, so long as the temperature is sufficiently low. At ordinary temperatures, thermal agitation must be taken into account—in which case a box full of earthworms would give a more realistic image. This material is soft and stretches irreversibly when pulled. If bridges are now created between the chains (as during vulcanization when sulfur bridges S–S are introduced between polyisoprene chains), a material is formed which keeps its shape: a *rubber*. Only a small concentration of bridges is required to make an "infinite" cluster of connected chains.

Branched polymers

Consider a solution of monomers all with the same functionality, f. If links are formed at all the extremities without any loops being made, then a Bethe tree is produced (see Fig. 3.1.3). The lack of available space is very quickly felt; a Bethe tree with excluded volume cannot be embedded in a finite dimensional space. Consequently, either the polymer contains a large number of *dangling* bonds, or else the polymer restructures itself by closing up at least these bonds.

A "Flory–Stockmayer" approach via a mean field theory, based on a Cayley tree model ($D = 4$) (see the mean field approach to percolation in Sec. 2.4.1), yields

$$N \propto R_0^4.$$

At the start of the 1980s, Isaacson, Lubenski, and de Gennes performed a calculation taking account of the excluded volume condition in a form similar to the case of linear polymers. The branching polymer is assumed to be in thermal equilibrium in a solvent in which it is sufficiently diluted. By minimizing the total free energy as shown above with respect to R, it is found that

$$\boxed{D = \frac{2(d + 2)}{5} \qquad (D = 2 \quad \text{when} \quad d = 3)} \qquad (3.5\text{-}9)$$

In fact, this calculation only corresponds to the fractal dimension if the system is restricted to a single cluster and to distances less than its size. In practice, when polymerization takes place, the medium becomes polydisperse, and this polydispersity must be taken into account when calculating the effective fractal dimension of the medium.

With a cluster distribution of the form $n_s = s^{-\tau}$ at the percolation threshold, the fractal dimension is found to be

$$\boxed{D_{\text{eff}} = D\,(3 - \tau)} \qquad (3.5\text{-}10)$$

so by taking $\tau \cong 2.2$, we have, for $d = 3$, $D_{\text{eff}} \cong 1.6$.

These two values, 2 for a monodisperse medium and 1.6 for a polydisperse medium, have been well verified experimentally.

It would also be interesting to investigate the structure of *molten polymers*, since then no solvent is present, bringing about a partial screening between the polymers, and so to an increase in the fractal dimension. The interested reader should consult specialist books and articles (de Gennes, 1979; Daoud and Martin in Avnir, 1989; etc.).

Gels

We must distinguish here between physical gels and chemical gels. With *physical gels*, we are dealing with polymers in fairly long chains in the presence of a "bridging agent" (such as Ca^{++}), unless they roll up in a spiral (Fig. 3.5.2). Bridges start to form between the chains at a concentration level determined by the temperature (the energy needed for these bridges to form is usually quite low). At high temperatures, there are few links and the polymer is liquid. At low temperatures, the system is frozen by the large number of links present. Between these two extremes, there is a gelation temperature at which the mean number of links present insures the existence of an infinite cluster. In all these cases the system can be maintained in a liquid or solid state by adjusting the temperature. Gelatine is a physical gel which melts at 60°C.

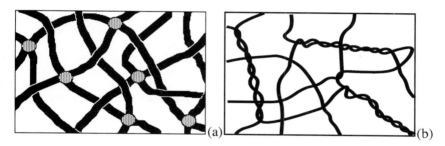

Fig. 3.5.2. Formation of a physical gel. Linear polymers may be bonded to each other by ionic bridges or roll up into a helix. Bridges may be formed in sufficient concentration to produce an infinite cluster and thus to form a gel. The temperature controls the opening of these bridges whose formation barrier is fairly low.

With *chemical gels*, the structure is constructed by chemical bonding between the constituents (epoxy resins and glues are examples). For a long time, the system does not reach equilibrium, because the aggregation involved is either of the diffusion or the reaction-limited cluster–cluster type (and is assumed to be relatively monodisperse). Because of their fractal structure, as the polymerized clusters grow, their mean concentration decreases and so they occupy more and more space. The monomer concentration of a cluster of radius R and mass $M = R^D$ is $C = M/R^d$, while outside the clusters it is nearly zero. When C has decreased sufficiently and reached the initial concentration C_0, the clusters end up touching each other [Fig.3.5.3 (b)]. Thus, their mean radius R_G satisfies

$$C_0 = R_G^{-(d-D)}.$$

The aggregation process produces a certain growth rate for the clusters, thereby causing their mean radius R(t) to depend on the time, t. Thus, at a particular time, t_G [$R(t_G) = R_G$], called the *gelation time,* an infinite cluster appears.

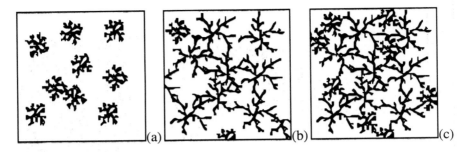

Fig. 3.5.3. Formation of a chemical gel: as long as the monomers are sufficiently dilute the polymerized clusters remain separate (a). At a time t_G, gelation occurs (b). After a longer time, there may be restructuration within the gel and densification (c). Polydispersity plays an important role in these processes.

The dynamics are, in fact, governed by the diffusion coefficient of each type of cluster. The larger their volume, the slower they diffuse. It is reasonable to assume a power law for these diffusion coefficients, $\mathcal{D}$ (M), a law which agrees with experimental results. Thus (Kolb and Herrmann, 1987),

$$\mathcal{D}\,(M) = M^\gamma \text{ with } \gamma < 0.$$

The growth rate depends essentially on γ. The temperature also plays an important role in these growth rates.

Flory's mean-field treatment of the successive branching of monomers using Cayley trees as models, although it gives perfect results for gels, ignores any loops and leads to steric constraints making it impossible for an infinite Cayley tree to be constructed in three-dimensional space. Observed deviations from this model led de Gennes and Stauffer in 1976 (see de Gennes, 1979, p.137) to propose a percolation approach to the case when no solvents are present. If a solvent is present, then, at the moment of gelation, the cluster–cluster process can no longer be diffusion-limited, as the clusters have already penetrated each other. The idea is that links form randomly between the clusters to produce a network similar to a percolation network. The number of these bonds continues to grow in time, in other words, the percolation threshold is exceeded (for more details see Jouhier et al., in Deutscher, Zallen, and Adler 1983, Chap. 8).

The passage from a sol (polymer solution) to a gel is easily observed due to alterations in the mechanical properties of the medium. At the moment the gel is formed, the viscosity of the sol increases and diverges, while from this point on the elasticity, which was nonexistent, begins to increase (see Sec. 5.2.4).

3.5.2 Fractal properties of membranes

Membranes are structures that play a crucial role in many areas of physics, chemistry, and biology. They are also conceptually very interesting because of their two-dimensional geometry ($d_T = 2$). Continuing in the same vein, we shall only examine those structural aspects of membranes that lead to fractal geometries. Depending on their local molecular constitution, membranes may present very different structures. On the one hand, there are fluid membranes whose constituent molecules diffuse easily through the heart of the membrane, and on the other, there are solid membranes obtained by polymerization with unbreakable molecular bonds.

Fluid membranes

Amphiphilic molecules are those which possess a hydrophilic (polar) head, along with one or more hydrophobic chains. These molecules readily situate themselves at the interface of two immiscible fluids such as oil and water; they are the surfactants used to make emulsions. They also commonly occur in a two-layered configuration in which the hydrophobic chains pair up. This is how the membranes of living cells are formed, with the hydrophilic heads on the outside in contact with the water of the cells. These membranes fluctuate thermically and, unlike solid membranes, have no resistance to tearing. Locally they can dilate like an elastic membrane.

A correlation length associated with these fluctuations can be defined. To do this we take the mean value of the scalar product of two unit normal vectors at two surface points at distance r apart. A plane surface corresponds to an infinite correlation length. The fluctuations cause the membrane to crumple and so, in principle, as the temperature increases we should expect a *crumpling transition*. Baumgärtner and Ho (1990) have recently carried out numerical

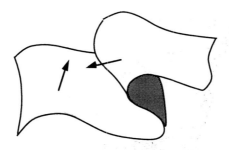

simulations of crumpling in fluid vesicles.[17] They found that the radius of gyration was related to the number of surface monomers by a power law of the form

$$N \propto R^D \quad \text{with} \quad D \cong 2.5 .$$
(3.5-11)

If we think of the crumpled fluid membrane as a self-avoiding random surface, this is not dissimilar to the external surface of a diffusion front as described in Sec. 3.3.1, whose fractal dimension was also close to 2.5.

---

[17] The calculations were carried out on vesicles rather than open membranes so as to avoid problems relating to the free boundaries.

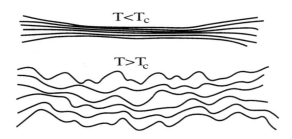

Fig. 3.5.4. Uncoupling transition in a pile of fluid membranes.

In addition, in a stack of membranes it is possible to observe an *uncoupling transition*. Between two membranes there exists an attractive van der Waals force which produces an interaction potential of the type shown in Fig. 3.4.2 (a). At low temperatures the membranes pile up in almost parallel layers. Above a certain critical temperature, the membranes separate (repulsive forces of entropic origin) due to the appearance of fluctuations in their geometries (Fig. 3.5.4). The uneven nature of this geometry does not appear to have been studied in detail.

Solid membranes

A solid or polymerized membrane is a surface which preserves the connectivity between each pair of points on its surface. An example of this is a surface composed of a set of interconnected particles (monomers) in a two-dimensional lattice. The scaling properties of such a surface depend neither on the structure of the lattice (triangular, square, etc.), nor on the precise nature of the short range particle interaction, but only on the fact that the bonds between the particles cannot be broken.

A crumpled piece of paper is a simple example of a solid membrane. Polymerized membranes can also be made from monomers interconnected along liquid–gas, liquid–liquid, or solid–gas interfaces. An example of this is poly(methyl-methacrylate) extracted from the surface of sodium montmorillonite clays. Polymerized membranes have also been obtained by copolymerization of lipidic bilayers.

In two dimensions, the analogues of solid membranes are the self-avoiding chains (polymerized chains moving along an interface) described in the previous paragraph. We have seen that they undergo a transition at the Θ-point.

In general, solid membranes involve a repulsive attraction between monomers, which tends to flatten them. Thus, they undergo a phase transition due to the competition between the forces tending to flatten the membrane (simulated by a coupling, J, between the normals at neighboring regions) and the entropy tending to crumple the surface. Fig. 3.5.5 shows four equilibrium configurations of a solid membrane for various values of $\kappa = J/kT$. The phase transition between the flat and crumpled phases occurs at a value close to

$\kappa = 0.5$.

The radius of gyration, R_g, of a membrane without excluded volume (no hard core interaction between monomers) in its crumpled phase, varies as $(\log R)^{1/2}$, where R is the linear size of the flat membrane. When the excluded volume interactions are present, the radius of gyration grows as R^{v_t}, where $v_t \cong 0.8$. You can easily check this exponent 0.8 by crumpling squares of paper with different length sides R. The crumpled phase is fractal and the mass–radius relation gives

$$\boxed{D = \frac{d-1}{v_t} \cong 2.5 \quad \text{in } d = 3} \quad . \tag{3.5-12}$$

Just as with linear polymers, we could make a mean-field calculation here by adopting the Flory approach; the expression obtained generalizes the calculation made for chains and we find

$$\boxed{D^{\text{Flory}} = \frac{d+2}{d_T+2} d_T} \quad . \tag{3.5-13}$$

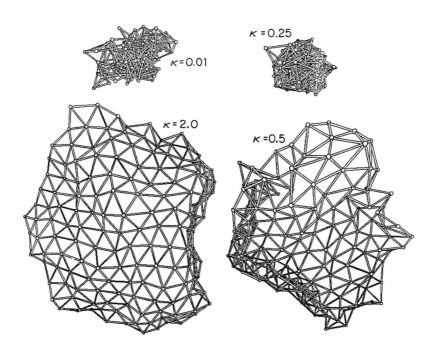

Fig. 3.5.5. *Equilibrium configurations of a hexagonal membrane composed of small triangular regions. They correspond to four different values of the reduced interaction,* κ, *between the normals at neighboring regions.* $\kappa = 0.01$ *and* $\kappa = 0.25$ *correspond to the crumpled phase,* $\kappa = 0.5$ *is close to the transition, and* $\kappa = 2.0$ *corresponds to the flat phase (Kantor, 1989).*

d_T is the topological dimension of the membrane or chain. For a membrane, D^{Flory} is exactly 2.5.

CHAPTER 4

Growth Models

Most of the progress that has been made in the study of aggregation phenomena in the last few years has derived from numerical simulations carried out on models of growth mechanisms. Dynamical features enter here due to the irreversible nature of these systems; time has a direction.

The complexity of the phenomena is reduced by supposing an elementary process to be the dominant mechanism, an assumption later confirmed. A simple model of cellular growth was proposed along these lines by Eden in 1961 to account for the tumor proliferation.

4.1 The Eden model

In its simplest version, this model constructs clusters on a lattice. As in the case of percolation, a cluster is composed of connected sites, two sites being connected if they are first neighbors. Cluster growth occurs by, at the end of each unit interval of time (taken as constant), an empty site, called a growth site, becoming occupied. In the Eden model, the growth sites, G, are those neighboring the cluster sites (i.e., the sites already occupied) (Fig. 4.1.1). The process begins with a germ (or seed, or nucleation site), which might, for example, be a line as shown in Fig. 4.1.2. Each new site occupied at time n is drawn at random from the growth sites G at time n−1.

Note that slight modifications to this site model (bond model) can easily be made, as they were to the percolation model.

When a (computer generated) Eden cluster is examined (see Fig. 4.1.3), it is found to be compact (except for a few holes close to the surface), but its surface is found to be rather tortuous.

The mass–radius relation in a d-dimensional Euclidean space takes the form

$$M \propto R^d ,$$

that is, its geometry is dense.

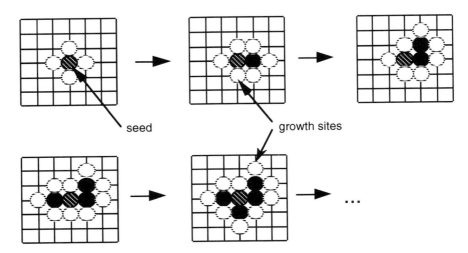

Fig. 4.1.1. Construction of an Eden cluster on a square lattice. The model simulates the proliferation of malignant cells (in black) starting from a single cell (the germ) by contamination of the closest neighbors (growth sites, in white). This contamination is assumed to occur at random within the set of growth sites.

Let us examine the surface structure in greater detail. To characterize this more easily it is helpful to choose instead other initial conditions, thereby avoiding the problems of radial geometry. The germ is taken to be the line $h = 0$ and periodic boundary conditions are chosen (column $x = L+1$ is identified with column $x = 0$). Growth can, of course, take place in higher $(d > 2)$ dimensions.

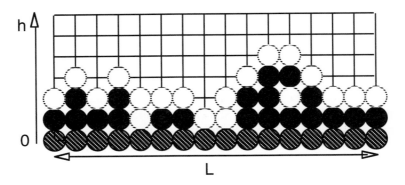

Fig. 4.1.2. Growth of an Eden cluster from a nucleation line (shaded sites).

Because of the global symmetry that is present, we know immediately that, if the surface structure is fractal, it has every chance of being a self-affine fractal, since the direction of growth, h, plays a very different role to the other directions (x, y, etc.), parallel to the line or surface of nucleation. The mean height of the cluster, $\langle h \rangle$, is simply the mean of the lengths h_i of the n_s points on its surface:

$$\boxed{\langle h \rangle = \frac{1}{n_s} \sum_i h_i} .$$

(4.1-1)

We could equally well take the mean height as $\overline{h} = N/L$, where N is the number of particles in the cluster, because, owing to the dense nature of the cluster, $\langle h \rangle$ and $\overline{h}$ very quickly converge to one another as $N \to \infty$.

Note that if the nucleation sites are excluded, N coincides with the time n (remember we assumed a constant rate of sticking here), and consequently both $\langle h \rangle$ and $\overline{h}$ vary linearly with time *for large enough values of N.*

Another important quantity (related to the extent of the surface's roughness) is the mean thickness of the surface defined by the variance of the height

$$\boxed{\sigma^2 = \frac{1}{n_s} \sum_i (h_i - \langle h \rangle)^2} .$$

(4.1-2)

4.1.1 Growth of the Eden cluster: scaling laws

When growth takes place from an initially flat surface, two types of behavior may be observed according to whether the mean height is greater or smaller than a critical height, $\langle h \rangle_c$, which depends on the width of the sample, L. The initially flat surface of the cluster becomes increasingly rough (σ increases with time, i.e., with $\langle h \rangle$). This continues until fluctuations in the value of h become sufficiently large in relation to the sample width, L, at which point these fluctuations group around a value dependent on L. An interesting aspect of these growth models is that they possess scaling laws, a property which we have found repeatedly in fractals, Brownian motion, fronts, etc.

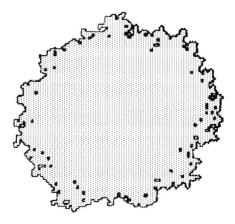

Fig. 4.1.3. Eden cluster composed of 1500 occupied sites generated from a seed (H.J. Herrmann, 1986).

Thus, the crossover point between régimes depends on L via a power law:

$$\langle h \rangle_c \propto L^z .\tag{4.1-3}$$

Which of the two regimes occurs depends on whether $\langle h \rangle$ is smaller or larger than $\langle h \rangle_c$:

$$\sigma(L,\langle h \rangle) \propto \langle h \rangle^\beta \quad \text{when } \langle h \rangle \ll L^z,\tag{4.1-4a}$$

$$\sigma(L,\langle h \rangle) \propto L^\alpha \quad \text{when } \langle h \rangle \gg L^z.\tag{4.1-4b}$$

Here, $\langle h \rangle_c$ plays the role of a characteristic length for the system, hence we should expect there to be scaling laws involving the ratio $\langle h \rangle / \langle h \rangle_c$. Numerical calculations confirm that for large enough values of h and L (presence of scaling law behavior), σ takes the general form

$$\boxed{\sigma(L,\langle h \rangle) \approx L^\alpha \ f\!\left(\frac{\langle h \rangle}{L^z}\right)} \ ;\tag{4.1-5}$$

the function f being such that

$$f(x) \propto x^\beta \quad \text{as } x \to 0, \text{ where } \beta = \alpha/z \text{ [to satisfy Eq. (4.1-4a)]}$$

and

$$f(x) \to \text{const.} \quad \text{as } x \to \infty \text{ [to satisfy Eq. (4.1-4b)]}.$$

Note that we could equally well express the evolution of the thickness, σ, in terms of a scaling law involving time: the system is flat at $t = 0$ ($\langle h \rangle = 0$), and since $\langle h \rangle$ is proportional to t,

$$\boxed{\sigma(L,t) \approx L^\alpha \ f\!\left(\frac{t}{\tau}\right)} \ ,\tag{4.1-6}$$

where $\tau \propto L^z$ is the characteristic time for a transitory régime to change to a stationary régime. These power laws and this scaling law, which not only involve the geometry of the system but also time, are called *dynamical scaling laws*.

Numerical results show that in dimension $d = 2$,

$$\alpha \cong 0.5,$$
$$\beta \cong 0.33,\tag{4.1-7}$$
and hence
$$z \cong 1.5.$$

Similar numerical calculations in $d > 2$ are difficult. This is due to the fact that the surface thickness, σ, is composed of two terms, one of these σ_{sc} (scaling) follows the scaling law above, the other σ_i is an quantity intrinsic to the model and is independent of L: $\sigma^2 = \sigma_{sc}^2 + \sigma_i^2$. This intrinsic contribution arises from the fact that new growth sites have the same chance of being occupied as the older ones; when $d > 2$ this prevents the values of the exponents α and β being found with any great accuracy. This effect may partially be eliminated by employing a *noise reduction* method. A counter set to zero is placed at each new growth site. Each time a site is chosen, its counter is incremented by one unit. When the counter reaches a predetermined value m, the site finally becomes occupied,[1] thus ensuring that the oldest growth sites have the highest probability of becoming occupied first. This reduces the intrinsic fluctuations, σ_i, without changing the exponents α and β (which are related to σ_{sc}) (see Fig. 4.1.4).

Fig. 4.1.4. Eden surfaces generated using three different values of the noise reduction parameter: from left to right, m=1, m=2, m= 4. This parameter corresponds to the number of times a site must be chosen for it to become occupied; the larger the value of m is, the more are the local fluctuations reduced (after Kertész and Wolf, 1988).

The results obtained by Wolf and Kertész in 1987 are as follows:

$$
\begin{aligned}
d = 2: \quad & \beta = 0.33 \pm 0.015, \quad & \alpha = 0.51 \pm 0.025, \\
d = 3: \quad & \beta = 0.22 \pm 0.02, \quad & \alpha = 0.33 \pm 0.01, \\
d = 4: \quad & \beta = 0.15 \pm 0.015, \quad & \alpha = 0.24 \pm 0.02.
\end{aligned}
\tag{4.1-8}
$$

These results are in good agreement with the conjecture, which is based on theoretical arguments, that $\alpha + z = \alpha + \alpha/\beta = 2$ (see Sec. 4.3).

We have seen that the body of a cluster produced in this way is *not* fractal. What can be said about the structure of its surface, inasmuch as it is known to exhibit power law behavior?

The self-affine character of the surface of an Eden cluster

At a given moment, the surface of an Eden cluster contains overhangs. That is to say, the height h is not a single-valued function of the abscissa r

---

[1] Thus creating new growth sites at which the counter is reset to zero.

$(0 \leq r \leq L)$ along the whole of the nucleation line (or of $\vec{r}$ in the case of a nucleation plane). It can, however, be numerically verified that these overhangs play no role in the scaling laws; in other words, the surface defined by the highest occupied sites [of ordinate $h(\vec{r}) = \max(h)$ at a given $\vec{r}$] possesses the same characteristics as the whole surface. Therefore, as regards those properties which relate to its large scale geometry, the surface may be represented by a single-valued function $h(\vec{r})$. If the surface is self-affine, we can calculate the auto-correlation function of the heights, as we did in Sec. 2.2.4, Eq. 2.2-8,

$$C(r) = \langle\, [h(\vec{r}_0 + \vec{r}) - h(\vec{r}_0)]^2 \,\rangle \,.$$

Thus, we should expect to find that

$$C(r) \propto r^{2H} \,, \tag{4.1-9}$$

where H is the Hurst exponent of the self-affine structure. Moreover, comparison with the previous results requires that $H = \alpha$. The thickness is found by calculating

$$\sigma^2 = \langle\, [\, h(\vec{r}_0) - \langle h(\vec{r}_0)\rangle\,]^2 \,\rangle = \langle\, [\, h(\vec{r}_0)]^2 \,\rangle - \langle h(\vec{r}_0)\rangle^2 \,,$$

and so, when the heights are uncorrelated, $C(r)$ becomes

$$C(r) = \langle\, [h(\vec{r}_0 + \vec{r}) - h(\vec{r}_0)]^2 \,\rangle = 2 \left(\langle\, [\, h(\vec{r}_0)]^2 \,\rangle - \langle h(\vec{r}_0)\rangle^2 \right) = 2\sigma^2 \,.$$

There is no correlation when r is sufficiently large (i.e., larger than a distance corresponding to the correlation parallel to the nucleation plane), which is the case when $r = L$. Hence (and this result has been well confirmed numerically),

$$C(L) = 2\sigma^2 \propto L^{2\alpha} \quad \text{et} \quad H = \alpha. \tag{4.1-10}$$

So, the surface of an Eden cluster is a self-affine structure, its local fractal dimension is given (cf. Sec. 2.2.4) by $D = d - \alpha$.

The Eden model is the simplest growth model. Several attempts to find an analytic solution have been made. Edwards and Wilkinson (1982) constructed a model which, by including nonlinear terms to take account of the irreversible nature of growth, achieved reasonably good values for the exponents, namely $\alpha = 1/2$ and $z = 3/2$. Subsequently, this equation was refined by Kardar, Parisi, and Zhang (1986); we shall return to this analytic approach when we come to look at deposition models in Sec. 4.3.

Certain features of the Eden model clearly demonstrate the difference that exists between an irreversible kinetic process and a critical phenomenon. For example, the lattice configuration may affect the structure: in a square lattice the cluster issuing from a nucleation site progressively acquires a diamond shape (however, this requires a large amount of computing time and more than 10^9 particles).

4.1.2 The Williams and Bjerknes model (1972)

This generalizes the Eden model by allowing a growth site neighboring a malignant cell to be contaminated with probability α and this malignant cell to become healthy with probability β.

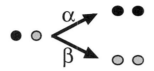

The important parameter here is the ratio $\kappa = \alpha/\beta$.

$\kappa = 1.1$

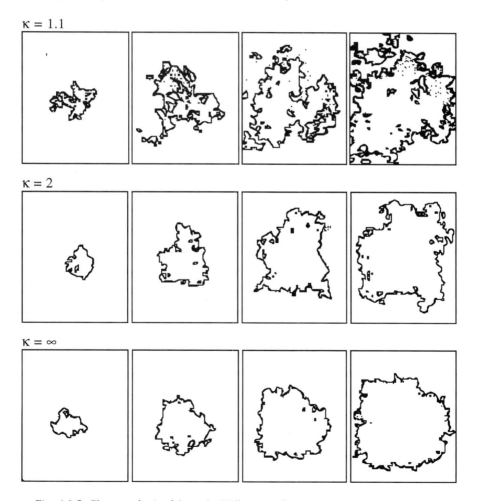

$\kappa = 2$

$\kappa = \infty$

Fig. 4.1.5. *Clusters obtained from the Williams and Bjerknes model for three values of the ratio α/β. The value $\kappa = \infty$ corresponds to the Eden model. (Williams and Bjerknes, 1972).*

If κ > 1, the contagion spreads. As with the Eden model, we find clusters of dimension d and self-affine surfaces (Fig. 4.1.5). There are very good reasons to think that this model belongs to the same universality class as the Eden model, and that, up to a scaling change (depending on κ), their clusters are similar.

4.1.3 Growth of percolation clusters

The Eden model may also be modified in the following way: a growth site G may be contaminated by a sick neighboring cell (occupied site of the cluster), but then only a proportion p of these sites G develop the illness and go on to contaminate their other neighbors, while the remaining $(1-p)$ become immunized and are no longer growth sites (Fig. 4.1.6). The case p = 1 corresponds to the Eden model.

This model is of great interest, because it is a growth model which generates percolation clusters. It is comparable to the invasion percolation model described in Sec. 3.2.3, but which could just as well have been included in this chapter. The two models are, in fact, different since, in invasion percolation, growth is never blocked and the proportion p of invaded sites progressively increases from zero until it approaches the value p_c; the invaded region thus becomes the incipient percolation cluster.

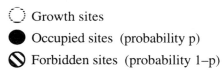

Growth sites

Occupied sites (probability p)

Forbidden sites (probability 1–p)

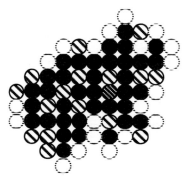

Fig. 4.1.6. Percolation cluster growing from a seed (heavily shaded); the cluster sites are shown in black, the forbidden (immunized) sites are shaded.

In the growth model examined here:

If $p < p_c$, the cluster develops until, at a certain moment, it reaches a finite size for which there are no more growth sites available (the perimeter being composed of immunized sites).

If $p > p_c$, an infinite cluster develops: the "contagion" has then spread a great distance.

When the growth stops, the cluster obtained is *identical* to a percolation cluster. However, *while the growth is in progress*, this is not at all the case: the cluster has the same fractal dimension as a percolation cluster, but globally it is less anisotropic, and its perimeter (growth sites) bears no relation to that of a percolation cluster.

4.2 The Witten and Sander model

4.2.1 Description of the DLA model

The Witten and Sander, or diffusion-limited aggregation (DLA), model is without doubt the most important particle–cluster aggregation model. It was proposed in 1981 by Witten and Sander and has since been the starting point for a great number of studies. The initial motivation was to explain the experiments of Forrest and Witten on the aggregation of smoke particles, but, as it turned out, the model applies much better to a different class of phenomena, as we shall see later. The model works as follows:

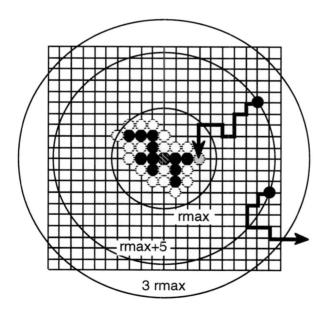

Fig. 4.2.1. Diagram of the DLA model on a square lattice. Growth takes place from a seed (hatched). The cluster sites are shown in black, the growth sites in white. The figure shows the trajectories of two particles, one of which drifts too far from the cluster and is lost.

Particle *n* diffuses through a lattice from a point of which, situated in an arbitrary direction far away from the developing cluster, it originates. When the particle arrives in contact with the cluster, it sticks *permanently*. Then particle *n+1* arrives, and so on. *In practice*, the numerical simulation is carried out following the scheme shown in Fig. 4.2.1: to prevent the computation time from being prohibitively long, the *n*th particle is positioned at random at a distance of the order of $r_{max} + 5$ from the cluster, where r_{max} is the distance of the furthest point from the initial germ. This has no effect on the future evolution of the system as the particle's *first encounter* with a circle not intersecting the cluster is equally likely to occur at any of its points. In addition, certain particles may wander too far from the cluster (and in $d \geq 3$ do so irreversibly). It may be assumed, but this is of course only an approximation, that the role of the departure point has sufficiently diminished at $r = 3 \, r_{max}$, and that it would waste too much computation time waiting for it to return towards the cluster (in three dimensions there is even a nonzero probability that we would wait forever!). When this happens we start again with a new particle. A random walk over the lattice simulates Brownian diffusion. Aggregates may also be constructed without the use of a lattice. Fig. 4.2.2 displays an aggregate from the Witten and Sander model (here of 50,000 particles) generated off-lattice.

Fig. 4.2.2. Example of an off-lattice DLA cluster [2] composed of 50,000 particles. The fractal dimension is found to be close to 1.71 (Feder, Fractals, 1988).

---

[2] In off-lattice diffusion, each particle moves with equal probability to any point one particle diameter away. When it touches a cluster particle during this motion, it sticks to the cluster.

In principle, the fractal dimension does not depend on the lattice but only on the Euclidean dimension d. By contrast, as with the Eden model, the global shape of the cluster is sensitive to the lattice: with a square lattice, the cluster takes on a shape with four petals, as is more clearly demonstrated when noise reduction is used (see, e.g., Meakin and Tolman in Pietronero, 1989). Whereas the fractal dimension is close to D = 1.715 in the off-lattice model, it appears to be close to 1.5 for a square lattice, which, as we might expect, introduces two exponents (longitudinal and transversal), as did the Eden model.

Table V below gives the values of the fractal dimension D, for $1 < d \leq 6$. No exact analytic expression for D(d) is known. Mean field arguments produce a good approximation to the fractal dimension

$$D(d) = (d^2 + 1)/(d + 1).$$

Table V: Numerically obtained fractal dimensions of the DLA cluster in various spatial dimensions (Meakin, 1983)

d	2	3	4	5	6
D	1.71	2.53	3.31	4.20	5.3
$(d^2+1)/(d+1)$	1.67	2.50	3.40	4.33	5.29

4.2.2 Extensions of the Witten–Sander model

Two types of extension are of physical interest. The first consists in allowing a probability of sticking other than 1. When the diffusing particle arrives in the neighborhood of a particle in the aggregate, there is now only a probability p < 1 of sticking taking place. The sticking, however, remains irreversible (when the particle crosses the sticking barrier it is trapped permanently). This model simulates a diffusion-reaction process.

In this model a *crossover* occurs. After a long time has elapsed and the cluster is reasonably large, the fractal dimension is independent of p and identical to that of the Witten and Sander aggregate just studied above. In contrast, at the start of the aggregation process, the cluster appears more compact. In particular as $p \to 0$, it takes on the appearance of an Eden model aggregate. It is easy to see the reason for this: as $p \to 0$ the particle visits a large number of sites neighboring the aggregate before it finally sticks. So, the probability of sticking to each of them is practically the same, and this approximately defines an Eden model. For any given value of p, after N of the sites neighboring the cluster have been visited, the probability of an (N + 1)th site being visited is no greater than $(1 - p)^N$. When the cluster has grown large, the sites with a non-negligible chance of being visited are limited to a small neighborhood of the point of first contact and we find ourselves in a scenario similar to the Witten and Sander model.

Another interesting extension consists in modifying the mean free path of the diffusing particle. In the Witten–Sander model, the mean free path is the distance between sites. In many aggregation processes (especially in gaseous phases), this mean free path, Λ, need not be negligible, and here again there is a crossover. If we suppose Λ to be infinitely large, then the regime is *ballistic*, the particles randomly describing directed straight lines in space. (This model was introduced by Vold in 1963.) Recent numerical calculations have given values for the fractal dimension of 1.93 and 1.95, and it is almost certain that in two dimensions the true value is 2. Similarly in three dimensions a value of 2.8 is found, but the true value is most probably 3: *the aggregates formed are compact*. For a finite value of Λ, we find locally compact aggregates up to a range Λ, then beyond this we pass to a Witten–Sander type structure. The purely ballistic case where the particles arrive from a single direction is a model for deposition; it generates rough, self-affine surfaces (see Sec. 4.3).

Many physical systems involving aggregation (which were mentioned above in the section on aggregation in a field) can be modeled by the DLA model. This is due to the fact that, roughly speaking, all of these systems turn out to be solutions to Laplace's equation.

Consider a particle diffusing in the DLA model before it sticks to the cluster. The probability, P, of finding the particle at position r is found by solving a diffusion equation:

$$\frac{\partial P(\vec{r},t)}{\partial t} = \mathcal{D}\,\Delta P(\vec{r},t) \ , \tag{4.2-1a}$$

where $\mathcal{D}$ is the diffusion coefficient. The boundary conditions are as follows:

— at a large enough distance (r_{ext}) from the cluster the particle has a uniform probability:

$$P(\vec{r}_{ext}, t) = P_0 \ ; \tag{4.2-1b}$$

— when the particle reaches the cluster it is adsorbed, and the probability of it penetrating inside the cluster is zero:

$$P(\vec{r} \in \text{cluster}, t) \equiv 0. \tag{4.2-1c}$$

If we take sufficiently long intervals of time to allow each particle to reach the cluster, then the aggregation process is approximately in a steady state, since the addition of a small number of particles to the cluster does not appreciably alter its structure. Therefore, to find the probability of a diffusing particle being at a given point, we have to solve (steady state)

$$\Delta P(\vec{r},t) = 0 \ , \tag{4.2-2}$$

with the boundary conditions above [Eqs. (4.2-1b) and (4.2-1c)].

If the more practical model shown in Fig. 4.2.1 is employed, the problem is more subtle, since, as the particles are eliminated at $3r_{max}$, the probability there is nil as it is in the cluster. If we use a uniform distribution of starting sites at $r_{max} + 5$ to calculate the probability distribution (which satisfies $\Delta P = 0$), we find that it reaches a maximum at an intermediary distance and that its equipotentials (P = constant) are still practically circular far from the cluster. By contrast, the starting sites at $r_{max} + 5$ are too close to the cluster to be equipotential. But viewed from a neighborhood of the cluster it appears as though the particles have originated from a uniform circular source situated far away.

The particle current is given by the gradient field, grad $P(\vec{r})$. As this gradient field is related to the local curvature of the surfaces of equal probability (equipotentials), the tips grow more rapidly than the fjords. We shall see that these remarks also apply to the examples in Sec. 3.4.5. First of all, it is important to stress here the role played by random fluctuations. Let us suppose that we solve Eqs. (4.2-1) by allowing the initially linear cluster to grow at a rate proportional to grad $P(\vec{r})$. The surface of the aggregate will then remain flat, but this situation is metastable and the least curvature will immediately develop a tip. Random fluctuations in the process of particles sticking one at a time to the cluster (DLA), or in the distribution of pore diameters or in the local probabilities of dielectric breakdown, thus play a crucial role in the growth of ramified fractal structures, by continually destabilizing the process and preventing it from being regular.

In this regard, it is instructive to see how the different wavelengths at the start of the process develop when the linear approximation is still valid. If we start with a structure with a small surface oscillation of wavelength λ, it will develop fairly quickly. In general there is a critical wavelength λ_c, such that for $\lambda > \lambda_c$ the wave front is unstable, while for smaller wavelengths it is damped. For instance, in the case of injection (viscous fingering), $\lambda_c \approx \sqrt{C_a}$ [Eq. 3.2-5], the fastest developing wave front having wavelength $\lambda_m = \sqrt{3}\,\lambda_c$.

Electrolytic deposition

Here, the circular anode is set at a potential V_0, the cathode (germ) being set at zero potential. Ions diffuse in the presence of an electric field which is stronger the nearer they approach a deposition point, where they will be reduced. Growth is thus more rapid in the vicinity of tips (Fig. 3.4.10). We can see immediately that when the conditions are such that the current is too weak for nonlinear or dynamic effects to occur, that is, when the ion concentration is low enough, electrolytic deposition is very similar to a DLA situation, provided that the sticking is immediate. However, this type of growth is still only very incompletely understood due to some complicated problems of electrochemistry. Moreover, the carrier fluid is not immobile but animated in the neighborhood of the points of hydrodynamical motion, and governed by a Navier–Stokes equation (see, e.g., Fleury et al., 1991).

Dielectric breakdown

Dielectric breakdown (Fig. 3.4.11) also belongs to the DLA family. To see this, we must look more closely at the growth model of this breakdown. Locally there is a discharge between neighboring molecules, which may be thought of as a cluster of low resistance bonds that close at each discharge (a plasma of ionized molecules is formed). Once a discharge has occurred between two molecules, the conductivity takes a higher value. New ruptures create links to the periphery. In an insulator, the Laplace equation is always satisfied, and the probability of discharge is proportional to the gradient at the discharge surface (as with DLA), or, more generally, to a *power of the gradient* there (DLA variant: Niemeyer, Pietronero, and Wiesmann model discussed below).

Injection into a porous medium

Here a fluid of low viscosity (i.e., with a large capillary number C_a, see Sec. 3.2.3) is injected into a porous medium. In experiments where a fluid is forced into another more viscous fluid, the *potential* represents the pressure, and the *field* represents the velocity field. For an incompressible fluid satisfying Darcy's law, the current, $\vec{\mathcal{V}}$ (flux per unit normal surface area, see Sec. 3.2.1), is proportional to the pressure gradient, P,

$$\vec{\mathcal{V}} = -\alpha \overrightarrow{\text{grad}}\, P. \tag{4.2-3}$$

The conservation constraint div $\vec{\mathcal{V}} = 0$ leads again to $\Delta P = 0$. The boundary conditions at the interface are only consistent with the DLA model when the viscosity of the injected fluid is negligible compared to that of the displaced fluid. In the experiment shown in Fig. 3.4.12, Daccord and his colleagues measured the dimension to be $D = 1.7 \pm 0.05$.

Laplacian fractals

The term "Laplacian fractals" (Lyklema, Evertsz, and Pietronero, 1986) is used to describe fractals that have been obtained from diffusion-limited aggregation processes, that is, those which model critical kinetic phenomena governed by the Laplace equation. The components are, as we have just seen, the Laplace equation together with a law indicating how the kinetics are related to this equation (Fick equation) and a random component. The Fick equation determines the probability current, while the random element (here the individual random motion of a particle) allows the fractal cluster to grow by constantly changing (at each sticking) the boundary conditions, in a process which is in fact very far from equilibrium. If we attentuate this random element by allowing a large number of particles to diffuse simultaneously, the Laplace equation will be satisfied but no fractal structure will be generated (more precisely, its development will be slowed down considerably, as when m is chosen very large in the noise reduction of Sec. 4.1.1). The kinetics of the diffusing particle's random walk may be subjected to a particular constraint to help better model a physical process, for example dielectric breakdown. Niemeyer et al. (1984) have studied the growth of Laplacian fractal structures for which the probability of growth satisfies the equation

$$P(i \rightarrow j) = \frac{|V(i) - V(j)|^{\eta}}{\sum\limits_{i,j} |V(i) - V(j)|^{\eta}} \quad , \tag{4.2-4}$$

i being a cluster site at potential $V(i) = 0$, j a growth site, and $P(i \rightarrow j)$ the probability (normalized to one) of occupying site j starting from site i.

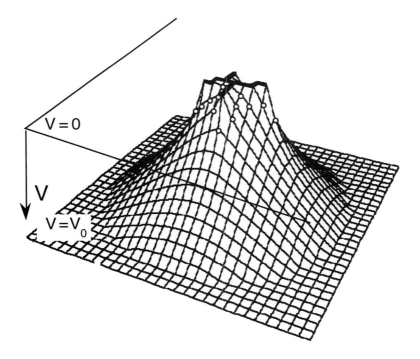

Fig. 4.2.3. *Diagram representing the potential between a circular electrode at potential V_0, ($V_0 > 0$), and an aggregate or an earthing point at potential $V = 0$ (indicated by a thick black line). Equivalently, this potential V represents the probability density P of a particle, originating at the boundary, being found at a given point. The probability current is given by Fick's law $\vec{J} = - \mathcal{D} \overrightarrow{grad} P$. The points adjacent to the cluster (white circles) are the growth sites. The probability of their being occupied is greater, the smaller is their ordinate (larger gradient), which is precisely the case for the tips of the cluster. In the case of electrical discharge, Niemeyer et al. have proposed that the probability of discharge between neighboring molecules is proportional to a power η of the electric field, that is,*
$$\vec{J} = - \mathcal{D} (\overrightarrow{grad} V)^{\eta}$$

The V(j)'s are calculated by solving the Laplace equation each time a new particle is added. The potential profile is shown in Fig. 4.2.3. The tips are in a stronger field and grow more quickly. For $\eta = 1$, we get a DLA cluster, although the microscopic process is different. Finally, for $\eta = 0$ we get an Eden cluster (equal probability for each growth site).

All these fractal structures are *self-affine*.[3] This is easier to see if the growth takes place from a line or a plane, as in Fig. 3.4.13 (see also Evertsz, 1990).

4.2.3 Harmonic measure and multifractality

During the growth of a DLA cluster, as we have already noticed, the growth sites are not occupied with the same probability. Growth sites situated on the tips of cluster branches have a higher than average weighting, while there are sites, situated at the ends of fjords formed by the branches, whose probability of being occupied is practically nil (Fig. 4.2.4). It is important to characterize this probability distribution, as it is directly involved in the growth of the cluster. Such a characterization does exist and uses multifractility.

In the case at hand, we give the following sketch of this multifractal approach. Suppose that we wish to investigate the charge distribution on a charged conductor of any shape placed in a vacuum. To do this we would have to solve the Laplace equation with boundary conditions $V = V_0$ at infinity and $V = 0$ on the conductor. For a surface element dS of charge dq, the surface charge density is $\sigma = dq/dS$, and the electric field, E_n, normal to the surface in the immediate neighborhood of the conductor is given by

$$E_n = \sigma/\varepsilon_0 . \tag{4.2-5}$$

The potential V_0 is also related to the charge distribution:

$$V_0 = \frac{1}{4\pi\varepsilon_0} \int \frac{dq}{r} . \tag{4.2-6}$$

The charge distribution on the surface of the conductor is called the *harmonic measure* (this name comes from the fact that Laplace's equation is satisfied by harmonic functions).

Thus, the problem of cluster growth is related to the notion of harmonic measure. For example, in the electrostatic problem associated with the DLA model, or with Laplacian growth, we have just seen that the probability $G(\vec{r})$ of a growth site being occupied, as a function of $\vec{r}$, is given by a power law in the electric field E_n and so also of σ ($\eta = 1$ for DLA):

$$G(\vec{r}) = \frac{E_n(\vec{r})^{\eta}}{\int_S d\vec{r}\, E_n(\vec{r})^{\eta}} . \tag{4.2-7}$$

---

[3] In the case of DLA, however, each branch appears to be self-similar.

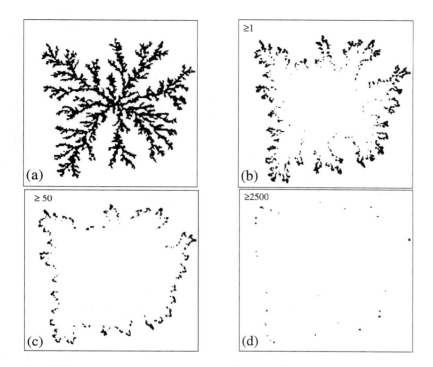

Fig. 4.2.4. Having constructed an off-lattice DLA cluster of 50,000 particles (a), we would like to know the probability distribution for the sticking of the 50,001th. To do this, a large number (10^6) of random walks are simulated and the number of times that the 50,001th diffusing particle comes into contact with each of the cluster particlest is determined. Figures (b), (c), and (d) show the particles which have been touched more than once, more than 50 times, and more than 2,500 times respectively (Meakin et al., 1986).

Unfortunately, it has not yet been possible to study the multifractal structure that corresponds to the harmonic measure for a DLA cluster analytically. If a is the size of a particle cluster, then (see Sec. 1.6, especially Sec. 1.6.4)

$$M_q \left(\varepsilon = \frac{b}{L} \right) \equiv \int_S d\vec{r} \; G \left(\frac{b}{L}, \vec{r} \right)^q = b^{1-q} \left(\frac{b}{L} \right)^{-\tau(q)}, \qquad (4.2\text{-}8)$$

where $b = \varepsilon L$ ($\varepsilon \ll 1$) is the size of the boxes centered at the different points $\vec{r}$, and $G(\varepsilon, \vec{r})$ is the cumulative probability (i.e., the measure inside a box). We can also fix b and vary the cluster size L to study the effect of increasing this size.

Knowing $\tau(q)$, the function $f(\alpha)$, which characterizes the multifractal structure, may then be found. This function achieves a maximum D_0 which is the dimension of the support of the measure, that is, the dimension of the DLA cluster itself: $D_0 \cong 1.71$. In their numerical calculations, Hayakawa et al. (1987) found $D_0 \cong 1.64$ and $D_1 \cong 1.04$.

> In two dimensions, a theorem of Makarov (1985) and Jones and Wolff (1988) states that the information dimension of the harmonic measure is exactly equal to one. This means that the fractal dimension of the support containing the majority of the charge is $D_1 = 1$, whatever the shape of the conductor (which is in good agreement with above). It has also been conjectured that in three dimensions the information dimension is equal to 2.5.

4.3 Modeling rough surfaces

We have already spoken implicitly (Sec. 3.4.1) about two growth models for rough surfaces: the Eden model (explained in Sec. 4.1) and ballistic aggregation. Of course, not all rough surfaces can be described by these two models alone; examples of those than cannot include surfaces generated by fractures, mentioned in Sec. 2.2.9.

4.3.1 Self-affine description of rough surfaces

The term "rough surface" defines an irregular surface in which there are no overhangs, or rather, in which *such overhangs do not dominate the scaling properties*. When these conditions are obtained, a rough surface may be correctly described by a function $h(\vec{r})$ which specifies the height[4] of the surface at position $\vec{r}(x,y)$ on an appropriate reference plane (in $d = 3$). For a mountain relief, the reference plane is mean sea level, and $h(\vec{r})$ is the altitude at the location $\vec{r}$ (defined by its latitude and longitude). For a great many rough surfaces occurring in nature, this function $h(\vec{r})$ is characteristic of a *self-affine fractal structure*. The growth of such a structure may thus be characterized by means of the parameters $\langle h \rangle$ and σ defined in the Eden model.

4.3.2 Deposition models

Deposition processes (from a vapor, by sedimentation, etc.) represent typical cases of the growth of rough surfaces. Clusters produced by deposition are dense, only their surfaces have an irregular structure. Various models have been studied in the last few years (a recent review can be found in Family, 1990, which we follow here). Here, we shall describe only three models: *random deposition, random deposition with diffusion,* and *ballistic deposition.*

Random deposition is the most elementary of these models. Here the particles fall vertically at random onto a horizontal substrate and pile up along

---

[4] If there are overhangs in the surface, we take $h(\vec{r}) = \max(h)$ at position $\vec{r}$.

each vertical line (Fig. 4.3.1). The heights $h(\vec{r})$ are thus uncorrelated and follow a Poisson distribution, so that whatever the value of d,

$$\langle h \rangle \propto t$$
$$\text{and} \quad \sigma \propto t^{1/2} . \tag{4.3-1}$$

The process does not depend on the sample size L.

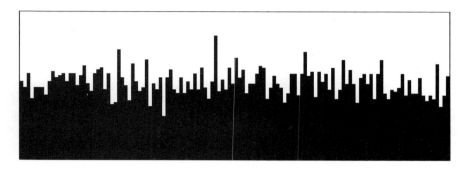

Fig. 4.3.1. Random deposition onto a one-dimensional substrate (d = 2) (Family, 1990).

Random deposition with surface diffusion is identical to the previous model except that instead of sticking immediately to the cluster, the particles "diffuse" before stabilizing at a more stable site situated within a given radius (the particles roll down into the holes) (Fig. 4.3.2). In this model restructurization is dominant; the rough surface generated now depends on the sample size. It behaves much like an Eden model, but belongs to a different universality class (i.e., its exponents have different values). In particular, the surface fluctuations obey a scaling law of the form

$$\boxed{\sigma(L,\langle h \rangle) \approx L^{\alpha} \ f\left(\langle h \rangle / L^{z}\right)} \quad , \tag{4.3-2}$$

where, as with the Eden model, the function f is such that

$$f(x) \to \text{const.} \quad \text{as } x \to \infty$$
$$\text{and} \quad f(x) \propto x^{\beta} \quad\quad \text{as } x \to 0, \ \text{where } \beta = \alpha/z. \tag{4.3-3a}$$

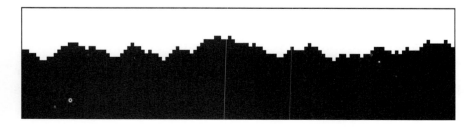

Fig. 4.3.2. Random deposition with diffusion (d = 2) (Family, 1990).

By adopting an analytic approach that uses Langevin's equation (Sec. 4.3.3), Edwards and Wilkinson (1982) have proposed the following relations:

$$\alpha = (3 - d)/2, \quad \beta = (3 - d)/4, \quad \text{and} \quad z = 2. \tag{4.3-3b}$$

The fractal dimension of this self-affine surface (with Hurst exponent $H \equiv \alpha$) is thus

$$D = d - \alpha = 3(d - 1)/2. \tag{4.3-3c}$$

Ballistic deposition: we have already spoken earlier about ballistic aggregation (Sec. 4.2.2). In this model introduced by Vold (1959), the direction in which the particles fall is normal to the substrate, and when they enter into contact with one of the cluster particles they stick there permanently (Fig. 4.3.3). Again the general behavior is very similar to that of the Eden model. Numerical studies lead to the following predictions concerning the scaling laws of the surface:

$$\sigma \propto t^{\beta}, \quad \text{where} \quad \beta \cong 1/3 \text{ in } d = 2. \tag{4.3-4a}$$

Finite size effects also occur,

$$\sigma \propto L^{\alpha}, \quad \text{where} \quad \alpha \cong 1/2 \text{ in } d = 2. \tag{4.3-4b}$$

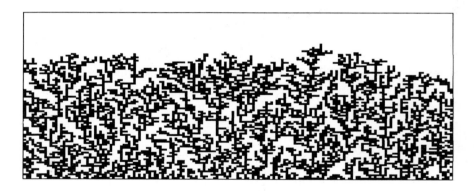

Fig. 4.3.3. Ballistic deposition along a direction normal to the substrate (Family, 1990).

There is no experimental reason for ballistic deposition to take place perpendicular to the substrate, and ballistic models involving an oblique incidence have been investigated. This leads to columnar structures similar to those observed experimentally. In this case, the exponents α and β vary continuously with the angle of incidence.

4.3.3 Analytic approach to the growth of rough surface

In 1986, Kardar, Parisi, and Zhang proposed an equation governing the variation in height $h(\vec{r},t)$ during the growth of a rough surface. This equation generalizes the approach developed by Edwards and Wilkinson in their study of random deposition with diffusion. It is a nonlinear Langevin equation,

$$\partial \tilde{h}(\vec{r},t)/\partial t = \gamma\, \Delta \tilde{h}(\vec{r},t) + \lambda \left(\overrightarrow{\mathrm{grad}}\ \tilde{h}(\vec{r},t)\right)^2 + \eta(\vec{r},t) \quad \text{with}\ \tilde{h} = h - \langle h \rangle \ .(4.3\text{-}4)$$

It contains a relaxation term whose coefficient is associated with the surface tension (as does Laplace's equation for an elastic membrane), a non-linear term corresponding to the lateral growth with coefficient λ (see Vicsek, 1989, p. 200 for further details) and a noise term, η, which allows the introduction of some randomness. In the models that we have just studied this noise is a white Gaussian noise satisfying

$$\langle\, \eta(\vec{r},t)\, \eta(\vec{r}\,',t')\,\rangle\ = 2C\, \delta(\vec{r} - \vec{r}\,')\, \delta(t - t')\ . \tag{4.3-5}$$

The case when $\gamma = \lambda = 0$ corresponds to random deposition with $\beta = 1/2$; if $\gamma \neq 0$, $\lambda = 0$, then we have deposition with restructurization, $\alpha = (3-d)/2$, $\beta = (3-d)/4$, and $z = 2$; finally, in the general case (ballistic deposition and the Eden model), we find for $d = 2$ that $\alpha = 1/2$, $\beta = 1/3$, and $z = 3/2$, which agrees with the numerical results obtained in $d = 2$. The situation remains unclear[5] in $d > 2$.

Let us end this section by adding that the Kardar, Parisi, and Zhang equation is also interesting for the study of other growth models where correlation phenomena intervene (molecular beam epitaxy, for example, can be modeled by random deposition process in which the deposited particles stabilize at sites so as to maximize the number of links, thus introducing a spatial correlation). Recent studies have therefore been led to a modified equation (4.3-4) in which the noise is no longer white but of the form

$$\langle\, \eta(\vec{r},t)\, \eta(\vec{r}',t')\,\rangle\ = 2C\, |\vec{r} - \vec{r}'|^{\alpha}\, |t - t'|^{\beta}\ . \tag{4.3-6}$$

4.4 Cluster–cluster aggregation

In Sec. 2.7.1 we introduced cluster–cluster processes and gave aerosols and colloids as experimental examples of these processes.

4.4.1 Diffusion-limited cluster–cluster aggregation

In 1983, Meakin and, simultaneously, Kolb, Jullien, and Botet studied the diffusion-limited cluster–cluster aggregation model numerically to characterize the structure of aggregates and the dynamics of their formation. The principle behind the simulation is simple. Initially, a low concentration of particles are randomly distributed over a lattice. The particles diffuse in time by

---

[5] Kardar et al. have shown, however, that for very large values of λ, it is always the case that $\alpha + z = 2$.

jumping to neighboring sites (double occupation of a site being forbidden) and permanently stick to each other when they touch. The pairs thus formed also randomly diffuse by translation through the inter-site distance, and the clusters stick as soon as they enter into contact with each other. Obviously, there is no reason why clusters of different sizes should have the same diffusion coefficient. For example, we could choose the diffusion coefficient for a cluster of size (mass) s to be

$$\mathcal{D}_s = c \, s^\gamma \, . \tag{4.4-1}$$

Thus, for D-dimensional clusters, taking $\gamma = -1/D$ leads to diffusion coefficients inversely proportional to their hydrodynamic radii (de Gennes, 1979), which for a fractal cluster is of the order of its radius (effective or of gyration).

This aggregation process may be easily generalized to the off-lattice case. As we can see from Table IV in Sec. 3.4.1, the aggregates formed in this way are less dense than those formed by the diffusion-limited particle–cluster process (DLA). This result does not depend on γ so long as $\gamma < 1$ (physically the most common case). When $\gamma \gg 1$, this model asymptotically approaches the DLA model, as there are then present only large clusters together with some small clusters that have escaped sticking. Although the large clusters diffuse very quickly here and the small ones remain almost motionless, the relative motion is similar to the DLA case (to convince yourself of this watch the film produced by Kolb, 1986).

The aggregation model can be refined by allowing the clusters to rotate. This does not qualitatively alter the results, or even the internal relaxations of these clusters, but it can lead to a certain restructuring making the clusters denser.

Scaling laws

The size distribution is an important parameter in the description of cluster–cluster aggregation phenomena. It is defined in terms of $n_s(t)$, the mean number of clusters of size s per site at time t. At a given time, the clusters have a mean size $\langle s \rangle(t)$. As with the previous examples involving a scaling régime, we would expect this mean size to obey a power law in time (see for example Vicsek, 1989),

$$\langle s \rangle(t) \propto t^z \tag{4.4-2}$$

and there to be be a scaling law of the form

$$n_s(t) \propto s^{-\theta} \, f(s/t^z) \, . \tag{4.4-3}$$

The exponent θ may be determined by fixing the mean concentration, yielding

$$\theta = 2 . \tag{4.4-4}$$

By contrast, both the value of z as well as that of the scaling function f depend on the exponent γ of Eq. (4.4.1) via the relation (Kolb in Stanley and Ostrowsky, 1985)

$$z = [1 - \gamma - (d - 2)/D]^{-1} . \tag{4.4-5}$$

• For $\gamma < \gamma_c$ the function f is a bell-shaped curve,

$$f(x) \propto x^2 g(x) \text{ where } g(x) \ll 1, \quad \text{if } x \gg 1 \text{ or } x \ll 1 \text{ [and } n_s(t) \text{ is then}$$
of the form t^{-2z}].

• For $\gamma > \gamma_c$ the function f is a monotone curve,

$$f(x) \ll 1, \quad \text{if } x \gg 1 \text{ (exponential cut of large sizes), and}$$
$$f(x) \propto x^\delta \text{ if } x \ll 1, \text{ and } n_s(t) \text{ is of the form } t^{-z\delta} s^{-2+\delta}.$$

The exponent δ is called the crossover exponent. Simulations in d = 3 give a critical value $\gamma_c \cong -0.27$, and numerical values of z and δ, as a function of γ, as given in the table below.

Table VI: Influence of diffusion [Eq. (4.4-1)] on growth

γ	−3	−2	−1	−1/D	−1/2	0	1/2
z	0.33	0.45	0.85	1	1.3	3	$\cong 100$
$2 - \delta$	–	–	–	–	$\cong 0$	1.3	1.87

4.4.2 Reaction-limited cluster–cluster aggregation

This model was created (Kolb and Jullien, 1984; Brown and Ball, 1985) to study the case where the sticking probability tends to zero (due to the presence of a potential barrier to be crossed before irreversible sticking takes place, as in the case of weakly screened colloids). The corresponding model for particle–cluster aggregation is the Eden model. Here again steric constraints due to the space occupied by each cluster lead to a lower fractal dimension than that of the dense Eden cluster (see the table in Sec. 3.4.1). Furthermore, the cluster dispersion plays a crucial role in reaction-limited cluster–cluster aggregation.

Cluster dispersion

In exactly hierarchical processes, the probability distribution of the cluster sizes peaks at a single value which varies in time: $n_s(t) = \delta[s - s_0(t)]$. In practice, monodispersity, of which the hierarchical model is an idealization, corresponds rather to a bell-shaped distribution centered on the mean cluster size. By contrast, in a polydisperse system, all sizes are present but with a predominance of small clusters (with the n_s distributed as s^{-a}).

In the case of reaction-limited cluster–cluster aggregation, monodisperse systems are less dense. The fractal dimensions found by numerical simulation (Kolb and Jullien, 1984) are:

$$D \cong 1.55 \quad (d = 2) \quad \text{and} \quad D \cong 2.00 \quad (d = 3) . \tag{4.4-6}$$

By contrast, polydispersity makes aggregates denser (Brown and Ball, 1984), the fractal dimensions then being

$$D \cong 1.59 \quad (d = 2) \quad \text{and} \quad D \cong 2.11 \quad (d = 3) . \tag{4.4-7}$$

These last values are in good agreement with the experimental results described earlier (e.g., the gold aggregates in Fig. 3.4.5).

The Smoluchowski equation

The mean variation of the cluster size may be investigated by means of a kinetic equation due to Smoluchowski, which is based on the aggregation speeds K_{ij} of clusters of different sizes i and j (making a cluster of size $i + j$). This may be written

$$dn_s(t)/dt = \frac{1}{2} \sum_{i+j=s} K_{ij}\, n_i(t)\, n_j(t) - n_s(t) \sum_{i=1}^{\infty} K_{is}\, n_i(t) . \tag{4.4-8}$$

Many studies have been conducted concerning the form of the solutions to this equation using various expressions for K_{ij} (Leyvraz 1984; Ernst, in Pietronero and Tossati 1986, p. 289; see also Vicsek's book 1989, and the article on aggregation by Jullien, 1986). This may also be generalized to the reversible case in which clusters once formed may split again into two clusters.

4.4.3 Ballistic cluster–cluster aggregation and other models

In this third type of cluster–cluster aggregation model, the clusters may move freely (or diffuse with a large mean free path compared to the cluster size). Aggregation may be purely hierarchical (the monodisperse case) or, instead, correspond to a polydisperse process. The fractal dimension does not appear to depend on the dispersity, and the values obtained (see Family and Landau, 1984) are

$$D \cong 1.55 \quad (d = 2) \quad \text{and} \quad D \cong 1.91 \quad (d = 3) . \tag{4.4-9}$$

These values are in agreement with measurements made on aerosols in a vacuum.

High density case

All the previous work has concerned situations in which the particle density is low. When this is the case, the clusters may effectively be considered as being situated on average reasonably far apart from one another. Highly dense systems behave differently. To try to understand the effect of high densities, in 1987, Kolb and Herrmann carried out a numerical study on a model in which the particle concentration was $\rho = 1$. The model was constructed as follows: all the lattice sites are initially occupied and bonds are created by making random attempts to displace the clusters already formed (starting with isolated particles) along vectors joining pairs of neighboring sites. Particles which enter into contact are then taken to be bonded, having returned to their initial positions.

In this model, surface or volume fractals are obtained depending on the value of γ. Thus, $\gamma = -2$ leads to surface fractals, $\gamma = 1$ to fractal aggregates, while when $\gamma = 2$ nonfractal objects are produced.

Reversible aggregation

All these modes of aggregation may be modified by allowing the possibility of reversible sticking of clusters or, more precisely, the possibility of clusters to divide in two. This type of process, which may also be investigated by means of Smoluchowski's equation, eventually leads to a steady state where the cluster size distribution is in equilibrium.

We shall not enter into further details concerning cluster–cluster aggregation as this is another very rich topic requiring a whole book to itself. Interested readers may consult the review articles relating to this subject or the book by Tamàs Vicsek dealing with growth phenomena (Vicsek, 1989 and 1991).

CHAPTER 5

Dynamical Aspects

Thus far, apart from occasional incursions into the realm of time-dependent phenomena, we have mainly been concerned with describing the geometric aspects of those fractal structures which are found in nature. These, however, have nearly always been the geometric aspects of a time-evolving system. Thus, we have successively examined ordinary and fractional Brownian motion (some of whose features are related to the present chapter), turbulent and chaotic phenomena, and the growth of aggregates whose dynamic scaling laws might also have been included in this chapter. Although these different aspects are interconnected, during this chapter, we shall devote ourselves to studying purely kinetic or dynamic characteristics.

5.1   Phonons and fractons

5.1.1 Spectral dimension

The question of spectral dimension is closely related to the problem of finding the density of vibration modes, or spectral density, in a fractal lattice. The spectral density is a very important quantity which is involved in problems concerning relaxation, transport, adsorption, etc.

Here we are interested in the component of the spectral density's behavior which is due to the lower frequencies. This corresponds to the long term behavior, that is, in the case of diffusion for example, to times that allow the explanation of the large scale structure of the system, which is sensitive to its fractal geometry, to be undertaken. In a d-dimensional Euclidean lattice this behavior is simple: the number of modes with frequency less than ω varies as ω^d :

$$N(\omega) \propto \omega^d \qquad\qquad (5.1\text{-}1)$$

so, the spectral density, defined by $dN(\omega) = \rho(\omega)\, d\omega$ is such that

$$\rho(\omega) \propto \omega^{d-1}, \qquad\qquad \omega{\to}0 . \qquad\qquad (5.1\text{-}2)$$

Is there an equally simple expression for a fractal? Shender (1976) and Dhar (1977) proposed a power-law behavior for a self-similar fractal,

$$\boxed{N(\omega) = \int_0^{\omega} \rho(\omega') \, d\omega' \quad \underset{\omega \to 0}{\propto} \quad \omega^{d_s}} \quad , \tag{5.1-3}$$

where d_s is called the *spectral dimension*.

Scaling law arguments have allowed Rammal and Toulouse (1983) to relate this exponent to the fractal dimension D and the scaling law satisfied by each separate mode.

Relation between the spectral and fractal dimensions

(i) Consider a fractal lattice composed of identical masses, linked by M identical elastic bonds in a cube of side L. Due to its fractal nature, change of scale by a factor b gives

$$M(bL) \propto b^D M(L),$$

so that the mode density (which depends on the number of elastic bonds) in a volume of side bL is

$$\rho_{bL}(\omega) \propto b^D \rho_L(\omega) .$$

(ii) As the modes are discrete and ω is positive (there is a ground state), let us make a bijection between modes. Then, *if there is a b* such that

$$\omega(bL) \propto b^{-\zeta} \omega(L),$$

we can deduce that

$$\rho_{bL}(\omega) \propto b^{\zeta} \rho_L(b^{\zeta}\omega) .$$

(We shall verify this hypothesis in the case of Sierpinski's gasket.)

From the two relations between ρ_{bL} and ρ_L, we have:

$$\rho_L(\omega) \propto b^{-D+\zeta} \rho_L(b^{\zeta}\omega) .$$

So for power-law behavior in ρ, it is sufficient to choose b such that $b^{\zeta}\omega = $ const.:

$$\boxed{\rho_L(\omega) \propto \omega^{d_s-1} \quad \text{with} \quad d_s = D/\zeta} \quad . \tag{5.1-4}$$

But what was the point of establishing this relation when we do not know ζ? In fact, as an example will show, ζ can be calculated — sometimes exactly, more often approximately — by a method of renormalization. This method is a practical illustration of the so-called renormalization group transformations (in this case in real space).

Example of calculation of the excitation spectrum by renormalization: the Sierpinski's gasket

Consider the gasket of side L, below, in which the vertices of the triangles of mass m are linked by springs of tension K. The coordinates x_1, x_2, x_3, y_1, y_2, y_3, X_1, X_2, X_3, etc., represent small displacements of the masses m about their equilibrium position. These displacements are vectors which may be restricted to the plane of the lattice if planar modes alone are being considered. The equations of motion are of the form

$$\text{m } d^2 \vec{x}_0 / dt^2 = \vec{F}_{01} + \vec{F}_{02} + \vec{F}_{03} \quad \text{with} \quad \vec{F}_{0i} = K \, (\, \vec{x}_i - \vec{x}_0 \,) \cdot \vec{n}_{0i} \; \vec{n}_{0i} \quad .$$

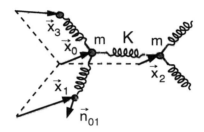

Such a system of equations is, however, too difficult to solve. To simplify it we carry out a *purely scalar calculation*, that is, using a single component per site. This is not a very realistic approximation for a network of springs (or a lattice of atoms, etc.), but gives us a first idea of the dynamic behavior of a fractal system. It may, however, be interpreted as a model of

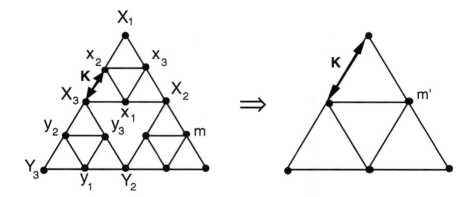

Fig. 5.1.1. *Renormalization of a Sierpinski gasket: the renormalization consists in finding an iteration n such that when it is applied to iteration n − 1 of the hierarchy, a system having the same physical characteristics is obtained. If the tension K in the springs is left unchanged, we must alter (renormalize) the masses (m → m') (see the text).*

entropic vibration,[1] the component x_i corresponding to a local entropy. Sierpinski's gasket then leads to a system of coupled, linear equations which can be solved by renormalization. Moreover, thanks to its linearity the system may be studied at a fixed frequency, ω.

The x, y... denote the "displacements" of the vertices of the smallest triangles (of side l); X, Y... the displacements of the vertices of the triangles of side l' = 2l, etc.

The equations of motion are

$$m\omega^2 x_1 = K\{(x_1 - x_2)+(x_1 - x_3)+(x_1 - X_3)+(x_1 - X_2)\},$$
$$m\omega^2 x_2 = K\{4x_2 - x_1 - x_3 - X_1 - X_3\}, \qquad\qquad (5.1\text{-}5)$$
$$m\omega^2 x_3 = K\{4x_3 - x_1 - x_2 - X_1 - X_2\}.$$

On eliminating x_3, these three equations give

$$x_1 = f_1 (X_1, X_2, X_3), \qquad x_2 = f_2 (X_1, X_2, X_3).$$

The same calculation is carried out for y_2, y_3 and the results brought together as

$$m\omega^2 X_3 = K [4X_3 - x_1 - x_2 - y_2 - y_3] .$$

We thus find an equation similar to Eq. (5.1-5) for X_3 as a function of $(X_1, X_2 ; Y_2, Y_3)$, but where the mass has been "renormalized":

$$m' = m [5 - m\omega^2/K] . \qquad\qquad (5.1\text{-}6)$$

The dimensionless parameters of the problem are $\alpha = m\omega^2/K$ and $\alpha' = m'\omega^2/K = m\omega'^2/K$ (the latter equality is obtained by conserving the mass while changing the frequency), α and α' being related by the recurrence relation

$$\alpha' = \alpha(5 - \alpha) . \qquad\qquad (5.1\text{-}7)$$

Now, instead of renormalizing the mass ($m \rightarrow m'$), we renormalize the frequencies ($\omega \rightarrow \omega'$) leaving the mass unchanged: in other words, the gasket remains unchanged on renormalization. For a change of scaling b = 1/2 (L = 4l $\rightarrow$ L = 2l') the square of the frequency is multiplied by 5 if, in (5.1-7), we neglect the term in α^2 in the neighborhood of $\omega = 0$. This gives the relation

$$(\omega_{L/2})^2 = 5 (\omega_L)^2 \quad \text{of the form} \quad \omega_{bL} = b^{-\zeta} \omega_L .$$

From this we can find ζ, then, using the relation (5.1-4) obtained for d_s,

$$\zeta = \frac{1}{2} \frac{\ln 5}{\ln 2} \quad \text{and} \quad d_s = \frac{D}{\zeta} = 2\frac{\ln 3}{\ln 5} \cong 1.364 . \qquad (5.1\text{-}8)$$

We have thus found the exponent d_s of the density of "scalar vibration" states of the Sierpinski gasket.

---

[1] The elasticity of the material then being, as with rubber for instance, of entropic origin.

The various dimensional exponents satisfy the inequalities $d_s \le D < d$. We shall return to the general case, where the displacements of the masses are represented by vectors, when we come to look at the elastic properties of disordered systems displaying a percolation structure. For the moment, this calculation of the spectral dimension will suffice for our needs.

> Note however that it is possible to construct fractals of finite ramification, generalizing the Sierpinski gasket, and for which the value of d_s may be calculated exactly. Hilfer and Blumen (*Fractals in Physics*, 1986, p. 33) have shown that the possible values of d_s are dense in the interval [1,2] for a suitable choice of two-dimensional fractal lattice. Furthermore, by enlarging the space by direct multiplication, structures can be constructed for which d_s is dense in the interval $[1,\infty]$.

Dispersion relations

This term denotes the relation between the wavelength of a vibration mode and its frequency. For normal elastic lattices this relation is linear: waves of wavelength Λ propagate at the speed of sound v, and the dispersion relation can be written simply as

$$\omega \propto v/\Lambda \ . \tag{5.1-9a}$$

To find the dispersion relation for the modes of vibration in fractal structures, let us consider a mode of "wavelength" Λ according to the d spatial directions. In fact we should not really talk about wavelength in this situation but instead about localization length, as sinusoidal waves do not propagate in such media. This mode defines vibration modes covering spatial domains V_Λ, of spatial extent Λ^d, and a number Λ^D of elastic bonds in the fractal lattice. It also has frequency ω. All other modes of lower frequency correspond to larger dimensions than Λ and for these modes the bonds in V_Λ may be taken as vibrating in phase. Those modes of higher frequency than ω and of lower wavelength than Λ, are invariant under changes in the boundary conditions of the domain V_Λ, so that the integrated spectrum of the low-frequency modes, absent from the localization domain V_Λ,

$$\Lambda^D \int_0^\omega \omega^{d_s - 1} d\omega$$

is independent of Λ and ω, because of the scale invariance of the fractal structure (see, e.g., Alexander, 1989). Hence, the *anomalous dispersion relation* for modes of vibration in fractal structures is,

$$\boxed{\omega \propto \Lambda^{-D/d_s}} \ . \tag{5.1-9b}$$

Localized modes of vibration associated with fractal structures are known as *fractons*. Experimental confirmation of the existence of fractons will be given in Sec. 5.1.4.

Lévy and Souillard (1987) have proposed that the form of the wave function of a fracton (i.e., the *mean* spatial distribution of its amplitude of vibration) be given by

$$\phi(\omega) \propto \frac{\Lambda^{D/2}}{r^{d-D}} \exp\left\{-\frac{1}{2}\left(\frac{r}{\Lambda}\right)^{d_\phi}\right\} \quad \text{where} \quad 1 < d_\phi < d_{min} . \quad (5.1\text{-}10)$$

(d_{min} is the chemical dimension or the dimension of connectivity, see Sec. 1.5.1). This function decreases rapidly with the distance r and has a range of the order of Λ.

In a recent study, Bunde et al. have suggested that fractons have a very multifractal character, and that the vibrational amplitudes at a given frequency and fixed distance from the center of mass of a *typical* fracton, may be characterized, as a function of the amplitude, by a logarithmic size distribution. It would then no longer be possible to define the localization length describing the exponential decrease of these amplitudes uniquely.

5.1.2 Diffusion and random walks

Let us now consider the problem of a particle diffusing through a fractal lattice. Let $P_i(t)$ be the probability of finding the particle at vertex i in a fractal lattice at time t. Let W_{ij} be the probability per unit time of a particle at site j jumping to site i. Balancing these probabilities gives the time evolution equation

$$\frac{dP_i(t)}{dt} = -\sum_j W_{ij}\{P_i(t) - P_j(t)\} . \quad (5.1\text{-}11)$$

To solve a set of coupled equations of this kind, we seek an expression for the probability $P(i,t\,|\,k,0)$ of finding the particle at site i at time t, given that it was at site k at time 0, by expanding it in terms of the eigenstates $\psi_i^{(\alpha)}$ of the system in the form,

$$P(i,t\,|\,k,0) = \text{Re} \sum_\alpha \psi_i^{(\alpha)*} \psi_k^{(\alpha)} e^{-z_\alpha t} \quad \text{with the initial condition} \quad P(i,0\,|\,k,0) = \delta_{i\,k} .$$

The eigenvalues of the evolution equation are such that

$$\sum_j T_{ij}\,\psi_j^{(\alpha)} = \lambda_\alpha\,\psi_i^{(\alpha)} , \quad \text{where} \quad T_{ij} = W_{ij} - \delta_{ij}\sum_h W_{ih} .$$

Comparing this with the evolution equation we find that $\lambda_\alpha \equiv -z_\alpha$ for diffusion, while $\lambda_\alpha \equiv -(\omega_\alpha)^2$ for the equivalent vibration problem, because of the second time derivative.

Knowing the P_i allows all the quantities describing the transport of this particle to be calculated. The mean square distance covered is then given by (the P_i being normalized: $\sum P_i = 1$)

$$\langle R^2(t)\rangle = \sum_i R_i^2\,P_i(t) . \quad (5.1\text{-}12)$$

Here again, since the lattice is self-similar, we may expect power-law behavior as a function of time. So, let us write

$$\boxed{\langle R^2(t)\rangle \propto t^{\,2/d_w}}\ .$$

(5.1-13)

where d_w is the fractal dimension of the walk with t taken as a measure (of the "mass"). d_w is as important an exponent as the fractal and spectral dimensions to which it is related. Let us suppose for simplicity that the W_{ij} are equal to W for the jumps between first neighbors, and zero otherwise. In a Euclidean lattice this leads simply to a Brownian walk whose behavior has already been shown to be:

$$\langle R^2(t)\rangle \propto t\ .$$

(5.1-14)

The path of the Brownian motion, t steps within a radius R, has a fractal dimension (refer to Sec. 2.2.1)

$$d_w = 2.$$

(5.1-15)

The situation for a fractal lattice is not the same; it is harder for the particle to diffuse and so we expect that $d_w > 2$.

Let us now take the Fourier transform of the evolution equation. Defining

$$\widetilde{P}_i(z) = \int_0^\infty e^{-zt}\, P_i(t)\, dt$$

(5.1-16)

we find

$$-z\,\widetilde{P}_i = \sum_{j=i+\delta} W\,(\widetilde{P}_j - \widetilde{P}_i),$$

(5.1-17)

$j = i + \delta$ denoting the first neighbors of i. We see that this equation is the same as Eq. (5.1-5), if the transformation $\omega^2 \leftrightarrow z$, concerning the (scalar) harmonic vibrations of the lattice is made. Hence, we may use all the arguments of the previous subsection.

Denote by $\rho_D(z)$ the density of modes (i.e., of the eigenvalues) of the diffusion equation. From above,

$$\int_0^z \rho_D(\omega')\, d\omega' \underset{z\to 0}{\propto} z^{\,d_s/2}\ .$$

(5.1-18)

In particular, this allows us to determine the probability that the random walk will return to the origin after a time t,

$$\boxed{P_{\text{return}}(t) = \int_0^\infty \rho_D(z)\, e^{-zt}\, dz \propto t^{\,-d_s/2}}\ .$$

(5.1-19)

To prove this relation, we take the probability of a return to the origin to be inversely proportional to the number of visited sites or equivalently to the mean probability of occupying a site, hence

$$P_{return}(t) = \langle P_i(t) \rangle = \frac{1}{N} \sum_{i=1}^{N} P_i(t) = \frac{1}{N} \sum_{\alpha=1}^{N} \exp(-z_\alpha t)$$

so, finally

$$P_{return}(t) = \int_0^\infty \rho_D(z) \, e^{-zt} \, dz \,.$$

We can use a scaling argument to determine the form (at large t) of $P(\vec{r},t)$. There is, in fact, a unique scaling length involved in this problem, namely the diffusion length

$$\lambda(t)^2 = \langle R^2(t) \rangle, \qquad \text{hence we may write}$$

$$P(\vec{r},t) \propto \lambda^{-D} f\left(\frac{r}{\lambda}\right), \tag{5.1-20}$$

where the coefficient λ^{-D} is imposed by the normalization of $P(\vec{r},t)$ over the fractal domain.

Therefore, $P(0,t) \propto \lambda^{-D}$, hence

$$\lambda \propto t^{d_s/2D}$$

and, since $\lambda \propto t^{1/d_w}$, we have : $\qquad P(\vec{r},t) \propto t^{-d_s/2} f\left(\frac{r}{t^{1/d_w}}\right)$,

and

$$\boxed{d_w = \frac{2D}{d_s}} \,. \tag{5.1-21}$$

In particular, we have $d_w = 2$ when $D = d_s (= d)$.

> In very general terms, the diffusion profile, that is, the probability density of random walkers in random systems, has a stretched exponential structure. i.e., the function f above has the form
> $$f(x) \propto \exp(-x^u), \text{ where } u = d_w/(d_w - 1).$$
> (See, e.g., Havlin and Bunde, 1989, on this point.)
> Furthermore, the anomalous transport of particles in disordered structures is sensitive to additional disorder due, for instance, to a nonuniform probability distribution for particles jumping from one site to another, or even to the existence of a bias (systematic external force acting on the particles) (Bunde in Stanley and Ostrowsky, 1988).

Link with modes of vibration in fractal structures

We have just established a relation relating the scaling lengths λ (mean distance covered) of physical phenomenon to their time scales via the density of states. This relation also applies to (scalar) vibrational systems, as the structure of the equations remains unchanged under the transformation $\omega^2 \leftrightarrow z$. This means that the characteristic length Λ of a fracton and its frequency (reciprocal of a time) are related by the relation

$$\Lambda \propto (\omega^2)^{-\,d_s/2D}.$$

The anomalous dispersion relation of the vibrational modes of elastic fractal structures is thus, as we described above [Eq. (5.1-9)],

$$\omega \propto \Lambda^{-D/d_s}.$$

5.1.3 Distinct sites visited by diffusion

First, the following remark needs to be made: the spectral dimension d_s characterizes the *mean* number $S(N)$ of *distinct* sites visited after N steps.

For Euclidean lattices this relation is exact:

$$S(N) \propto \begin{cases} N^{1/2} & \text{when} \quad d = 1 \\ N/\ln N & \text{when} \quad d = 2 \end{cases} \Bigg\} \text{ the walk is recurrent.}$$
$$\qquad\quad N \quad \text{when} \quad d \geq 3, \text{ there is a nonzero chance to escape to infinity.}$$

For a fractal lattice, the number of distinct sites visited is proportional to R^D, where R is the mean radius of the walk, hence *via* Eqs. (5.1-13) and (5.1-21),

$$S(N) \propto N^{d_s/2} \quad \text{if} \quad d_s \leq 2, \tag{5.1-22}$$

and obviously no more than N sites may be visited after N steps,

$$S(N) \propto N \qquad \text{if} \quad d_s > 2. \tag{5.1-23}$$

From this we conclude that $d_s = 2$ is a *critical spectral dimension*[2] for random walks, that is to say

(i) if $d_s \leq 2$ the walk is recurrent and the exploration is called *compact*;

(ii) if $d_s > 2$ the walk is transitory and the majority of sites will never be visited, so the exploration is *noncompact*.

Considerations about the number of distinct sites visited during a diffusion process are important for many physical phenomena. An example of this is provided by diffusion in a trap possessing medium.

Diffusion in the presence of traps

Consider particles (or excitations), diffusing in a medium until they meet a trap which either annihilates or stops them for a very long time. How does the particle or excitation density evolve with time when this medium is a fractal structure? This is an important feature in numerous problems of charge transport, mainly in semiconductors (disordered alloys).

---

[2] Not to be confused with the dimension of the walk itself!

The time evolution can be described by calculating the *probability of survival* $\phi(N)$ of a particle after N steps,

$$\phi(N) = \langle (1 - x)^{R(N)} \rangle,$$

where $R(N)$ is the number of distinct sites visited and x is the density of traps; its mean, whose behavior we have just examined, is given exactly by $S(N) = \langle R(N) \rangle$.

At relatively short time scales,

$$\phi(N) \propto \exp[-xS(N)], \qquad (5.1\text{-}24)$$

a behavior which has been verified experimentally to a high degree of precision.

At relatively long time scales, $\phi(N)$ can be shown to behave theoretically like a "stretched" exponential, that is, as

$$\phi(N) \propto \exp\left(-c\{xS(N)\}^{2/(d_s+2)}\right). \qquad (5.1\text{-}25)$$

This may be explained by the presence of large regions without traps which are rarer the larger they are. In practice, except when d = 1, this behavior is very difficult to observe experimentally. Note that the problem of diffusion with traps, in a homogeneous medium, can be solved exactly and that the above results are in good agreement with the homogeneous case if d_s is simply changed to d (Havlin in Avnir, 1989).

5.1.4 Phonons and fractons in real systems

Localized vibration modes can be observed in real systems. Investigation by means of light and neutron scattering has demonstrated the existence of fractons and their essential properties.

The clearest experiments have been carried out on silica aerogels, but other experiments have used epoxy resins, solid electrolytic glasses of silver borate, and hydrogenized amorphous silicon. The experiments are tricky and the results not always convincing, so it is indispensible to have a wide range of evidence from different techniques. The case of aerogels, for which there are numerous measurements, covering a wide range of techniques, is discussed in detail. Polymers and gels are introduced in Sec. 3.5.1.

We have already had occasion to talk about silica gels when we were describing the measurement of the fractal dimension of aggregates (small angle scattering in Sec. 3.4.2). In very schematic terms, aerogels are prepared in the following manner: compounds $Si(OR)_4$ are hydrolyzed in the presence of alcohol diluting the product to form groups $\equiv Si-O-H$. These groups "polycondense" to form siloxane bridges,

$$\equiv Si-O-H + \equiv Si-O-H \rightarrow \equiv Si-O-Si \equiv + H_2O,$$

thereby producing a solid lattice based on SiO_4 tetrahedra. Aggregates grow around the nucleation sites until they stick to each other, leading to the construction of percolation clusters (cf. Fig. 3.5.3). Thus, a gel is formed. With this method of production the final product is called an alcogel as the solvent is made up of alcohols R–OH issuing from the initial product. The alcogel is then dried[3] to remove the interstitial fluid, yielding an *aerogel*.

To simplify matters, let us suppose that up to a distance ξ the final gel is a fractal of dimension D, and that it is homogeneous at longer distances. This is tantamount to bonds continuing to form between the aggregates after the gelation point has been passed. Knowing D and the density $\rho = \rho(\xi)$ of the material, ξ may be found:

$$\rho(\xi)/\rho(a) = (\xi/a)^{D-d} \quad (d = 3),$$

where a is the size of the grains (the compact aggregates at the basis of the fractal structure) and $\rho(a) \cong 2{,}000$ kg/m³, close to the density of pure silica.

The phonon–fracton transition

At distances large compared with ξ, in other words, for low-frequency vibrations, the fractal behavior plays no part as the fractal domains vibrate as one. The gels then behave like a normal solid elastic of dimension L, in which phonons are propagated at the speed of sound[4] v, with the dispersion relation $\omega \propto v/\lambda$, and density of states $\rho_L(\omega) \propto L^d \omega^{d-1} = L^3 \omega^2$.

At short distances, the fractal nature should bring about the existence of localized fractons, with the dispersion law $\omega \propto \lambda^{-D/d_s}$ and density of states $\rho_L(\omega) \propto L^D \omega^{d_s-1}$.

The fracton and phonon régimes meet when λ is of the order of ξ. This defines a crossover frequency $v_{cr} = (\omega_{cr}/2\pi)$ which depends on the mean density ρ. Thus,

$$\omega_{cr} \propto \rho^{D/d_s(3-D)}. \tag{5.1-26}$$

As mentioned above, the experimental observation of fractons is tricky (see, e.g., Courtens et al. in Pietronero, 1989, p. 285).

Light scattering experiments (Brillouin scattering, Raman scattering) have enabled Pelous and his collaborators at Montpellier and Courtens at IBM-Zurich to display the existence of fractons and to determine their essential characteristics. Courtens et al. (1988) have estimated the fractal and spectral dimensions to be $D \cong 2.4$ and $d_s \cong 1.3$, respectively. Furthermore, they have found the crossover frequency to be $\omega_{cr} \propto \rho^{2.97}$. Vacher et al. (1990) have measured the density of vibrational states in silica aerogels by a combination of inelastic, incoherent neutron spectroscopy and Brillouin scattering measurements. These measurements clearly demonstrate an extended fracton regime.

---

[3] A hypercritical drying process is used to avoid liquid–vapor interface tensions which tend to cause the deterioration of the fragile structure of the skeleton of siloxanes.

[4] Here, only consider acoustic phonons, propagating by compression, are considered.

Several authors have also tried to exhibit fracton-type behavior in amorphous materials (such as hydrogenized amorphous silicon), in certain epoxy resins, in ionic superconducting silver borate glasses, and at the interfaces of sodium colloids in NaCl. In the diluted antiferromagnetic composite $Mn_{0.5}Zn_{0.5}Fe_2$, Birgeneau and Uemura (1987) observed a crossover from a magnon régime change to a fracton régime (characterized by a very wide band).

For the correct interpretation of experiments on propagation and localization in these materials, a thorough understanding of their structure is important; this, however, is far from being the case. The structure of aerogels, for example, is still subject to debate: a percolation-type structure of elementary entities (SiO_4 or mini-aggregates), an aggregation-type structure and a hierarchical sponge-like structure have all been proposed (Maynard, 1989). Similarly, fracton modes are not clearly understood. Certain measurements suggest an interpretation using the scalar fracton model (these we calculated in Sec. 5.1.1 on a Sierpinski gasket), others seem rather to correspond to vectorial fractons. Several fracton branches (corresponding to compression and flexion modes) were observed by Vacher et al. (1990), who used a combination of Raman and neutron spectroscopy to display a succession of régimes in the vibrational dynamics of two aerogels with different microstructures. In this way, three crossovers were observed: at low frequencies, the change from a phonon to a fracton régime; at high frequencies, the change from fracton to particle modes (Porod régime in Q^{-4}); while at intermediate frequencies, there is a crossover, attributed by the authors, to the change from compression to flexion modes. The values obtained for the spectral dimension (and for the fractal dimension) depend on the microstructure. They found $D \cong 2.4$, $d_s(\text{flexion}) \cong 1.3$, and $d_s(\text{compression}) \cong 2.2$ in one case, and $D \cong 2.2$, $d_s(\text{flexion}) \leq 1$, and $d_s(\text{compression}) \cong 1.7$ in the other. These values suggest that internal connections occur more frequently here than in a three-dimensional percolation lattice (the tensorial elasticity of a percolation lattice is treated in Webman and Grest, 1985). Sec. 5.2.4, on the viscoelastic transition, may also be consulted on this subject.

5.2 Transport and dielectric properties

5.2.1 Conduction through a fractal

For an (ordinary) Euclidean object the conductance is given by a simple power of its transverse dimensions. For a uniform conductor in the shape of a cube of side L with two opposite faces conducting, the conductance G(L) of this object varies as L^{d-2}.

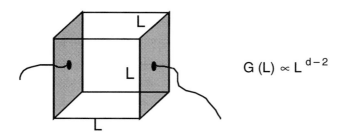

$$G\,(L) \propto L^{\,d-2}$$

In addition, the conductivity, Σ,[6] and the diffusion coefficient, $\mathcal{D}$, are related by the Nernst–Einstein equation

$$\Sigma = \frac{ne^2}{kT}\,\mathcal{D} \qquad (5.2\text{-}1)$$

(n is the concentration of charges e) and the diffusion coefficient can be calculated from $\langle r^2 \rangle = 2\,\mathcal{D}t$.

For a fractal, the Nernst–Einstein equation still holds for an effective coefficient of diffusion $\mathcal{D}_{\text{eff}}\,(r)$, defined by

$$\mathcal{D}_{\text{eff}}\,(r) = \frac{1}{2}\,\frac{d\langle r^2 \rangle}{dt} \qquad (5.2\text{-}2)$$

from Sec. 5.1.2:

$$\mathcal{D}_{\text{eff}}\,(r) \propto t^{\,d_s/D\,-\,1} \quad, \qquad (5.2\text{-}3)$$

that is, in terms of the mean distance $r = \sqrt{\langle r^2 \rangle} \propto t^{\,d_s/2D}$:

$$\boxed{\mathcal{D}_{\text{eff}}\,(r) = r^{\,-\theta} \quad \text{with} \quad \theta = 2\left(\frac{D}{d_s} - 1\right)} \qquad (5.2\text{-}4)$$

The conductance may also be determined:

$$G(r) \propto \sigma(r)\,r^{D-2} \propto r^{D\,(1\,-\,2/d_s)} \quad. \qquad (5.2\text{-}5)$$

Thus, conductance for a fractal structure depends anomalously on the dimension of the sample via a power law:

$$\boxed{G(r) \propto r^{\,-\tilde{\zeta}} \quad \text{where} \quad \tilde{\zeta} = \frac{D}{d_s}(2 - d_s)} \quad. \qquad (5.2\text{-}6)$$

It is interesting to compare this behavior with the results of Sec. 5.1.2. If $d_s < 2$ the walk is recurrent (the same points are passed indefinitely) and in this case the conductance decreases with the distance r: a localization phenomenon is produced. In particular, vibrations do not propagate when $d_s < 2$, but instead remain localized within certain regions of the fractal structure.

---

[6] The conductivity Σ relates the current density to the electric field $i = \Sigma\,E$, while the conductance G relates the current to the applied potential through $I = G(L)\,V$.

By way of an example, let us calculate explicitly the conductance of a Sierpinski gasket. As in the case of vibrations, renormalization is used for these calculations. Each set of three triangles (with a corner pointing upwards), whose sides have resistance R, is replaced by a single triangle with sides of resistance R':

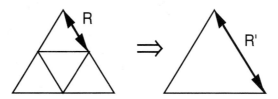

The resistance can be calculated exactly by taking the symmetry into account, and by noticing that the following two networks, with a current I flowing through them are electrically equivalent,

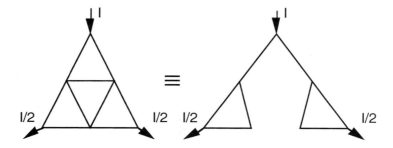

hence R' = 5/3 R. Therefore, at the nth iteration the conductance (which is in 1/R) is given by

$$\begin{cases} G^{(n)} = \left(\frac{3}{5}\right)^n G_0 \\ L^{(n)} = 2^n \end{cases} \quad \text{thus} \quad \frac{\ln G}{\ln L}_{n \to \infty} = \frac{\ln (3/5)}{\ln 2} = D\left(1 - \frac{2}{d_s}\right)$$

giving $d_s = 2$ (ln 3/ln 5). The result obtained for the Sierpinski gasket in Sec. 5.1.1 by numbering the vibration modes, is retrieved.

Comparison of the different approaches

It is instructive to compare approaches which deal with vibration in a fractal lattice with those dealing with diffusion or conduction in the same lattice. In the scalar case, the elastic energy can be written as

$$\frac{1}{2}\sum_{i,j} K_{ij} (x_i - x_j)^2,$$

where K_{ij} is the elastic constant between i and j. (All the K_{ij} with i and j nearest neighbors were taken equal to K in the example in Sec. 5.1.1.)

Equilibrium for these forces is found by minimizing this energy, or alternatively by putting $\omega = 0$ in the equations of motion [Eq. (5.1-5)]. Thus,

$$\sum_j K_{ij}(x_i - x_j) = 0. \tag{5.2-7}$$

If x_i is replaced by the potential v_i at site i and K_{ij} by the conductance Y_{ij} of the bond ij, this equation is identical to the Kirchoff equation for the same network.

Therefore, *the elastic constant of a network of springs behaves in the same way as the conductivity of an identical resistor network.*

5.2.2 Conduction in disordered media

One problem commonly found when dealing with disordered materials is that of determining the conductivity of a random mixture of two components of widely differing conductivities σ_1 and σ_2 ($\sigma_2 \gg \sigma_1$). An excellent review of this is in Clerc et al. (1990).

The conductivity depends on the relative concentrations of the two materials. Not surprisingly, the conductivity changes rapidly at the percolation threshold where a crossover takes place.

To help understand this phenomenon the following limiting cases may be considered:

(i) metal–insulator transition, corresponding to $\sigma_1 = 0$, σ_2 finite;

(ii) metal–superconductor transition, corresponding to σ_1 finite, σ_2 infinite.

In the first case, the conductivity varies just as it did for percolation (Fig. 3.1.2). Let p denote the concentration of the more highly conducting material (of conductivity σ_2). The material is an insulator when $p < p_c$; above p_c, the conductivity increases, and close to p_c it may be written (at least in a large network, neglecting the effects of finite size) as

$$\Sigma = 0, \quad p < p_c; \quad \Sigma \approx \sigma_2 (p - p_c)^{\mu}, \quad p \ge p_c \tag{5.2-8}$$

For percolation, the exponent μ (some authors use the notation t) is about 1.3 in two dimensions (see the table of dynamic exponents below).

In the second case, the conductivity of the material is finite when $p < p_c$; above the threshold, when percolation takes place, a superconducting path appears and the conductivity becomes infinite. Thus, we have

$$\Sigma \approx \sigma_1 (p_c - p)^{-s}, \quad p < p_c; \quad \Sigma = \infty, \quad p \ge p_c. \tag{5.2-9}$$

μ and s are critical transport exponents.

A network involving a metal–insulator transition is known as a *Random Resistor Network* (RRN). The corresponding diffusion problem has been around for a long time (de Gennes, 1976). A particle is constrained to diffuse through a percolation cluster. This situation is usually called the problem of the *ant in a labyrinth.*

A network involving a metal–superconductor transition is known as a *Random Superconductor Network* (RSN). The corresponding diffusion problem, in which a particle diffuses normally from one cluster to another, then infinitely quickly through each cluster, is usually (with a physicist's humor) referred to as the problem of the *termite in a disordered network.*

General case

From these limiting cases and the assumption that scaling laws exist, the general behavior for a mixture of two finite conductivities σ_1 and σ_2 can be deduced (Straley, 1976). The two cases link up in the neighborhood of p_c where the critical behavior occurs.

A reasonable hypothesis now is to take Σ as dependent essentially on the ratio σ_1/σ_2 and on $p - p_c$:

$$\Sigma\,(p,\,\sigma_1,\,\sigma_2\,) \approx \sigma_1\ f\!\left(\frac{\sigma_1}{\sigma_2},\,p - p_c\right). \qquad (5.2\text{-}10)$$

Then we assume the existence of a scaling law, that is, a dimensionless quantity[7] dependent on σ_1/σ_2 and $p - p_c$. Suppose that dividing the ratio σ_1/σ_2 by a factor b, corresponds to a dilatation of $(p - p_c)$ by a factor b^λ. Then $(\sigma_2/\sigma_1)^\lambda\,(p - p_c)$ is a dimensionless quantity and the conductivity Σ has the general form

$$\Sigma \approx \sigma'\left(\frac{\sigma_2}{\sigma_1}\right)^u\ g\!\left((p - p_c)\left(\frac{\sigma_2}{\sigma_1}\right)^\lambda\right). \qquad (5.2\text{-}11)$$

The prefactor with exponent u gives the scaling behavior at $p = p_c$; λ and u remain to be determined.

Finally, we take Σ to be nonsingular for $p = p_c$ and adjust the general behavior to agree with the limiting behavior. Good limiting laws therefore require that when

$$v > 0\ , \qquad\qquad g(v) \propto v^{\,\mu}$$

and that when

$$v < 0\ , \qquad\qquad g(v) \propto v^{\,-s}$$

so that:

(i) when $p > p_c$ and $v \gg 0$,

$$\Sigma \approx \sigma_1\left(\frac{\sigma_2}{\sigma_1}\right)^{u + \lambda\mu} (p - p_c\,)^\mu \approx \sigma_2\,(p - p_c\,)^\mu$$

must be independent of σ_1 , hence

$$\lambda = \frac{1 - u}{\mu}\ ;$$

(ii) when $p < p_c$ and $v \ll 0$,

---

[7] That is, in this case dilation invariant.

$$\Sigma \approx \sigma_1 \left(\frac{\sigma_2}{\sigma_1}\right)^{u-\lambda s} (p - p_c)^{-s} \approx \sigma_1 (p - p_c)^{-s}$$

must be independent of σ_1, hence

$$\lambda = \frac{u}{s}.$$

Finally, we have (see Fig. 5.2.1)

$$\Sigma \approx \sigma_1 \left(\frac{\sigma_2}{\sigma_1}\right)^{\frac{s}{s+\mu}} g\left((p - p_c) \left(\frac{\sigma_2}{\sigma_1}\right)^{\frac{1}{s+\mu}}\right) \qquad (5.2\text{-}12)$$

This expression is very interesting as it allows us to generalize the case where the component resistences (bonds of the disordered network) are replaced, using analytic continuation, by complex impedances.

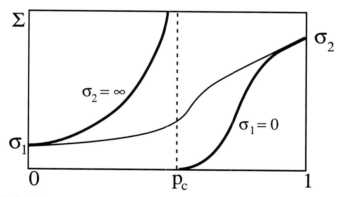

Fig. 5.2.1. *Diagram showing the variation of conductivity as a function of the concentration p of the more conductive bonds in the network ($\sigma_2 \gg \sigma_1$). The curve $\sigma_1 = 0$ corresponds to the insulator–metal transition, the curve $\sigma_2 = \infty$ to the metal–superconductor transition. The intermediate curve represents the general case where both conductivities take finite values.*

Particular case: at the percolation threshold $p = p_c$,

$$\Sigma \approx (\sigma_1)^{\frac{\mu}{s+\mu}} (\sigma_2)^{\frac{s}{s+\mu}}. \qquad (5.2\text{-}13)$$

Thus, in two dimensions the duality between the networks (Bergmann, in Deutscher et al., 1983, Chap. 13; see also Clerc et al., 1990) gives $s = \mu$ and an exact result:

$$\Sigma(p = p_c) = \sqrt{\sigma_1 \sigma_2}. \qquad (5.2\text{-}14)$$

Relation between μ and the spectral dimension d_s

Consider a random walk over the infinite percolation cluster when $p > p_c$. Two régimes are then observed: at short distances the medium appears to be a fractal of dimension D and this remains so up to a distance equal (on average) to the correlation length ξ. The correlation function is approximately of the form

$$g(r) \cong \frac{1}{r^{d-D}} \, \exp\left(-\frac{r}{\xi}\right) \, . \tag{5.2-15}$$

Thus, when $\quad \langle R^2(t) \rangle < \xi^2 \quad$ (that is, $t < t_\xi$) , $\quad \langle R^2(t) \rangle \propto t^{d_s/D}$.

On the other hand, for long times, when the diffusion distance $\sqrt{\langle R^2 \rangle}$ is greater than ξ, the medium appears homogeneous. Then the diffusion coefficient $\mathcal{D}$, which is time-independent, depends on $(p - p_c)$ as a power law whose exponent ρ,

$$\boxed{\langle R^2(t) \rangle \propto t \; \mathcal{D}(p) \propto t \; (p - p_c)^\rho} \tag{5.2-16}$$

will be determined by linking up these two régimes:

$$\begin{cases} \xi^2 \propto (t_\xi)^{d_s/D} \propto t_\xi \cdot (p - p_c)^\rho \\[2mm] \text{with} \quad \xi^2 \propto (p - p_c)^{-2\nu} \end{cases}$$

Identifying these two behaviors in the neighborhood of p_c and expressing t_ξ as a function of $(p - p_c)$ gives

$$\boxed{\frac{d_s}{D} = \frac{2\nu}{2\nu + \rho}} \, . \tag{5.2-17}$$

Now, we introduce the Nernst–Einstein relation $\Sigma(p) \propto \mathcal{D}_0(p)$, where $\mathcal{D}_0(p)$ corresponds to the diffusibility averaged over all sites. As only those particles placed in the infinite cluster contribute to $\langle R^2 \rangle$ as $t \to \infty$, the diffusion coefficient must be normalized relative to the probability $P(p)$ that the departure point belongs to the infinite cluster. Thus,

$$\mathcal{D}(p) = \frac{\mathcal{D}_0(p)}{P(p)} \propto \frac{\Sigma(p)}{P(p)}$$

or $\quad (p - p_c)^\rho \propto (p - p_c)^{\mu - \beta} \quad$ implying that $\quad \rho = \mu - \beta$. $\tag{5.2-18}$

So, *in the case of percolation*, we obtain the following relation between μ and d_s:

$$\boxed{d_s = \frac{2(d\nu - \beta)}{\mu - \beta + 2\nu} = \frac{2D}{\mu/\nu + D + 2 - d}} \, . \tag{5.2-19}$$

The question has often been raised of whether or not the dynamic exponents are related to the static (geometric) exponents. In 1982, Alexander and Orbach conjectured that for percolation $d_s = 4/3$ independently of d, which implies, at least for percolation, a relationship between the geometry and dynamics via the equation $d_s = 2D/d_w$ [and hence $d_w = (3/2)D$]. This result is certainly exact for $d \geq d_c = 6$ (the critical dimension for which the mean field is correct), and in numerical studies d_s remains very close to this value 4/3 (this would lead to a value $d_w = 2.84375$ for $d = 2$ compared with the value 2.87 obtained by numerical simulation). However, the perturbation expansion in $\varepsilon = d_c - d$ ($\varepsilon \to 0$) contradicts this supposition as it gives $d_s = 4/3 + c\varepsilon + ...$. In fact, no rigorous proof exists for d, an integer (i.e., $\varepsilon = 1, 2, ...$), and the numerical confirmation is tricky as d_w must have a precision of the order of 10^{-3}.

Table VII: Table of some dynamic exponents for percolation

d	d_w	$d\,^{\ell}_w$	d_c	μ/n	s/v
2	2.87	2.56	1.31	0.97	0.97
3	3.80	2.80	1.33	2.3	0.85

In the above table we have also shown the values of that exponent, $d\,^{\ell}_w = d_w / d_{min}$, which refers to those walks corresponding to the shortest path or chemical distance ℓ between two points (see Sec. 1.5.1 and Havlin, 1987):

$$\langle \, \ell(t) \, \rangle \propto \ell^{1/d^{\ell}_w} . \tag{5.2-20}$$

Other physical problems, very similar to that of conduction, have been studied in a similar framework. Examples of these are:
— the dielectric constant of an insulator–metal mixture;
— the critical current in a diluted superconductor;
— the flexion modes of a holed plate close to the percolation threshold (see, e.g., Wu et al. 1987); and
— the fracture energy of a composite material.
In Sec. 5.2.3 the dielectric behavior of a composite medium will be examined in greater detail.

Random resistor network: The Coniglio–Stanley model

The critical behavior of the conductivity during the metal–insulator transition of a random resistor network (composed of bonds of resistance R) has been discussed. To find out more about how the current is distributed through the network, especially when the bonds only support a certain maximum current, it is important to have an idea of the connections present in the network. A percolation cluster close to the threshold may be imagined as follows (see also the end of Sec. 3.1.1): the cluster is composed of a *backbone* onto which branches that do not participate in the conduction, because they are dead ends, are grafted (*"white"* bonds). The backbone itself may be partitioned into new subsets: *"red"* bonds, such that cutting any one of them would break the current in the system (all the intensity is found in these simply connected[8] bonds); and multiconnected "*blue*" bonds which constitute what is known as the "*blobs*" of the infinite cluster (Fig. 5.2.2). Within a region of side L, 1 << L << ξ, the number of bonds in each category follows power laws (Coniglio, 1981, 1982; Stanley and Coniglio, 1983, and references therein),

$$N_{red} \propto L^{D_{red}},$$
$$N_{blue} \propto L^{D_B},$$
$$N_{white} \propto L^D. \tag{5.2-21}$$

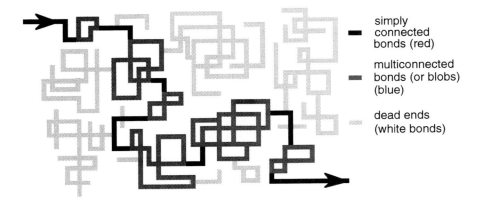

Fig. 5.2.2. *Schematic representation of a region of size L (L < ξ) of a percolation cluster (p > p_c) with current flow (in the direction of the arrows). The finite (not connected) clusters are not shown. A "red bond" is a relative notion which depends on the size of window chosen. These simply connected bonds disappear when L > ξ at the expense of the multiply connected bonds which are uniformly distributed.*

The following properties may be observed. The white bonds, through which no current flows, make up the bulk of the cluster, their fractal dimension being D (the same as the infinite cluster). The subset through which the current passes has dimension D_B (see Sec. 3.1.1); the blue bonds are dominant in this and have the same dimension D_B. Finally, the red bonds comprise a very tenuous yet very important subset. They exist (on average) only in domains of size less than or of the order of ξ, the connectivity length. Coniglio (1982) has shown that the red bonds have dimension $D_{red} = 1/\nu$, where ν is the exponent of the percolation correlation length.

Consequently, if overload is feared in a disordered medium, the further one is from the percolation threshold (p > p_c) the weaker is the current in the

---

[8] A bond is simply connected when its removal is sufficient to interrupt the current (this notion depends on L). A part of a cluster is simply connected if the removal of just one bond is sufficient to disconnect it.

red bonds of domain size ξ, since the current will then spread throughout these different domains.

The approach just described implies that the random resistor network has a multifractal structure. The multifractal measure to take is the current flowing through each bond.

Multifractal measure in a random resistor network

From what we have just seen, the red (simply connected) bonds carrying all the current correspond in the multifractal distribution to the subset with the maximum intensity $I = I_{max}$, while the white bonds through which no current flows correspond to the minimum intensity $(I = 0)$. Between these two extremes, the multiconnected bonds carry currents of all sizes.

The current distribution in the network (if the maximum current is fixed, $I_{max} = 1$) is then completely determined by the moments

$$M_q^I \equiv \langle (I/I_{max})^q \rangle = \langle I^q \rangle = \sum_I n(I)\, I^q \quad . \tag{5.2-22}$$

The potential distribution for a fixed current is identical (to within a factor R^{-q}) since $V = RI$ in each bond. The resistances of the bonds are taken to be one $(R = 1)$. We can also think of this distribution with the total potential fixed $(V_{max} = 1)$,

$$M_q^V \equiv \langle (V/V_{max})^q \rangle = \langle V^q \rangle = \sum_V n(V)\, V^q \quad . \tag{5.2-23}$$

The behavior is different in this case because I_{max} and V_{max} are related by $I_{max} = G(L)V_{max}$, which depends on the size of the sample:

$$M_q^V = G^q M_q^I \quad . \tag{5.2-24}$$

Some moments have a simple interpretation: M_0 is the mean number of bonds in the backbone, M_2^I is the mean resistance (for a fixed total current), and M_2^V is the mean conductance (for a given total voltage). Indeed the mean power dissipated per bond is $R\langle I^2 \rangle = \langle V^2 \rangle/R$, and for simplicity both R and either the total current or voltage are taken as unity. Similarly M_4 can be shown to be related to the amplitude of the noise present in the network.

The coefficient of hydrodynamic dispersion can also be related to the moment of negative order, $q = -1$. Indeed, if we consider a liquid flowing through a porous medium, then the pressure and speed of flow are related by an Ohm's law type of equation. If a colored marker (tracer) is injected through the entry face of the porous medium for a short time, it will disperse in the course of time and the dispersion will be given by the value $\langle t^2 \rangle - \langle t \rangle^2$, where t is the transit time of a colored particle. For each channel (bond) this time is inversely proportional to the current I, and hence $t^q \leftrightarrow I^{-q}$. Furthermore, the probability of a colored particle being present at the entrance to a channel is proportional to I. Therefore calculating the moment $\langle t^q \rangle$ is equivalent to finding M_{1-q}. Thus, the dispersion of the tracer is related to M_{-1}.

From what we know about the multifractal formalism (Sec. 1.6) the moments M_q obey power laws in the relative size ε of the balls covering the multifractal set. As before (see, e.g., Sec. 1.4.1), it is preferable to vary the size L of the network rather than $\varepsilon = b/L$, b being the length of a bond. Thus,

$$M_q \left(\frac{b}{L}\right) \propto \left(\frac{b}{L}\right)^{-\tau(q)} \doteq \left(\frac{b}{L}\right)^{(q-1)D_q} . \tag{5.2-25}$$

Certain values of τ are thus directly related to known exponents. The exponents, τ^V, at a fixed potential, and, τ^I, at a fixed current are related via the equation

$$\tau^V(q) = \tau^I(q) - q\tilde{\zeta} . \tag{5.2-26}$$

[Eq.(5.2-24) and the scaling behavior of G, Sec. 5.2.1, have been used].

The first values of τ are thus :
$\tau(0) = D_B$: the backbone dimension; $\tau^I(2) = \zeta$: the resistance exponent (Sec. 5.2.1); $\tau^I(\infty) = 1/v$: the dimension of the set of red bonds.

The results of numerical simulations giving the distribution of potential and of $\tau^V(q)$ for a given applied voltage are shown in Fig. 5.2.3 (de Arcangélis et al., 1986).

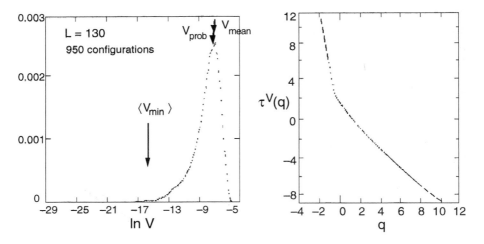

Fig. 5.2.3. Distribution of voltages in a random resistor network at the percolation threshold (figure on the left). The positions of the most probable and mean voltages are indicated. The figure on the right represents $\tau^V(q)$. The high voltage domain roughly obeys a scaling law in L^{-1}, while the low voltage one is in $L^{-6.5}$ (de Arcangelis et al., 1986).

Hierarchical models, conduction in the infinite cluster

Due to the difficulty of carrying out analytic calculations on the infinite percolation cluster and its backbone, various approximate models have been studied with the aim of understanding the behavior of transport in disordered media. Examples of these are the hierarchical models of Mandelbrot-Given (1984, see Fig. 1.4.6) and of de Arcangélis et al. (1986), in which the generator is

which gives on iteration two:

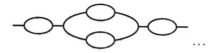

Besides allowing analytic calculations, these models have the advantage that their structure is fairly similar to that of the infinite cluster backbone. They have "*blobs*" and red and blue bonds forming subsets whose fractal dimension can be calculated exactly.

Transport in the case of continuous percolation

Continuous percolation, which models a certain number of physical phenomena in where the empty spaces or forbidden regions correspond approximately to randomly distributed spherical zones, exhibits different transport properties (diffusion, electrical conduction, elasticity, permeability) from percolation in a discrete network.[9]

This is due to the special behavior shown by the distribution of the "strength" $g(\delta)$ (conductivity, modulus of elasticity, etc.) of each bond (Fig. 5.2.4), where δ is the thickness of material between two empty regions.

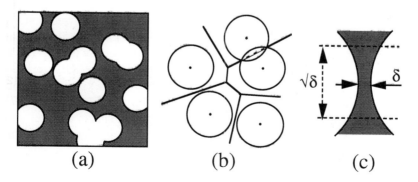

(a)   (b)   (c)

Fig. 5.2.4. Random distribution of empty regions in a material. (a) Model of continuous percolation (here d = 2) where the matter is shown in grey. (b) Associated network of bonds. (c) Parameter δ characterizing the "strength" of a bond (its conductivity, modulus of elasticity, etc.)

---

[9] Although the geometric critical exponents are the same.

Halperin, Feng, and Seng (1985) demonstrated that by using a uniform random distribution for the empty spheres in a random bond percolation lattice [Fig. 5.2.4(b)], the conductivity distribution g obeys a power law:

$$p(g) \propto g^{-\alpha}.$$

For small values of δ, which are the only ones playing a role in the critical behavior of transport in the neighborhood of the percolation threshold p_c, the "strength" of a bond varies as a function of δ, as $g(\delta) \propto \delta^m$ where m is given by the following table:

"Strength"	d = 2	d = 3
Diffusion	m = 1/2	m = 3/2
Constant of flexion force	m = 5/2	m = 7/2
Permeability		m = 7/2

The distribution of g is thus given by $p(g) = h(\delta) \, (dg/d\delta)^{-1} \propto g^{-\alpha}$ where $\alpha = 1 - 1/m$, since from the distribution of δ, $h(\delta) \to$ const., as $\delta \to 0$.

If we consider the conductance of the system, we find $\alpha = -1$ in d = 2, and $\alpha = 1/3$ in d = 3. The distribution p(g) is singular in three dimensions. The conductivity is given by Eq. (5.2-8) with μ now depending on α. Straley (1982) has argued for an exponent $\mu \, (\alpha)$ such that

$$\mu(\alpha) = \min [\, \mu, \mu_1 \, (\alpha)]$$

with
$$\mu_1 \, (\alpha) = 1 + (d - 2)\nu + \frac{\alpha}{1 - \alpha}$$

μ is the usual percolation exponent over a lattice (Sec. 5.2.2), and $\mu_1 \, (\alpha)$ is the exponent obtained by taking the conductivity to be limited by the "red bonds" (Fig. 5.2.2) of the associated random network (Halperin et al., 1985). We can therefore see that a crossover exists for a value α_c determined by $\mu = \mu_1$. ($\alpha_c \cong 0.23$ in d = 2, $\alpha_c \cong 0.13$ in d = 3). This behavior has been numerically verified by Bunde et al. (1986). Benguigui (1986) carried out experiments on conductivity and elasticity in copper sheets, and Lobb and Forrester (1987) have undertaken similar experiments for randomly holed sheets of steel and molybdenum.[10] The exponents obtained are in very good agreement with the theoretical model. Finally, we should add that the diffusion exponents for continuous percolation depend on the manner in which the diffusion takes place; a random walk of fixed step length gives different exponents (Petersen et al., 1989). For further details the work of Bunde and Havlin (1991) may also be consulted.

---

[10] In order to minimize the plastic deformations (copper and aluminium are more ductile).

5.2.3 Dielectric behavior of a composite medium

The results obtained in the previous paragraph are valid in a more general setting. They may justifiably be continued analytically and frequency dependent effects may be introduced because the scaling law has no singularity.

An example of this occurs when one of the components of a mixture is a *dielectric*. A mixture of a metal of conductivity σ_M and a dielectric of permittivity ε_I may be studied by introducing the complex conductivity of the dielectric

$$\sigma_I = \frac{\varepsilon_I\, \omega}{4\pi i} \; . \tag{5.2-27}$$

If we put $z = \sigma_I/\sigma_M$, the preceding results apply on substituting $\sigma_1/\sigma_2 \rightarrow \sigma_I/\sigma_M$, if $|z| \ll 1$, and $\sigma_1/\sigma_2 \rightarrow \sigma_M/\sigma_I$, if $|z| \gg 1$. In particular, the (complex) conductivity of the medium is given by the expression (Efros and Shklovskii, 1976; Bergmann and Imry, 1977)

$$\Sigma = \text{Re}(\Sigma) + \frac{\varepsilon\, \omega}{4\pi i} \; , \tag{5.2-28}$$

where, p_M being the concentration of the metal M,

$$\varepsilon \; \propto \; \begin{cases} \varepsilon_I \left| p_M - p_c \right|^{-s} & \text{when} \;\; |z| \ll 1 \\[2ex] (\varepsilon_I)^{\frac{\mu}{\mu+s}} \left(\dfrac{\sigma_M}{\omega}\right)^{\frac{s}{\mu+s}} & \text{when} \;\; |z| \gg 1 \end{cases} \tag{5.2-29}$$

$$\text{Re}\,(\Sigma) \; \propto \; \left. \begin{cases} \sigma_M\,(p_c - p_M)^{-s}, & p_M < p_c \\[2ex] \dfrac{\omega^2\,\sigma_I^2}{\sigma_M}\,(p_M - p_c)^{-\mu - 2s}, & p_M > p_c \\[2ex] (\sigma_M)^{\frac{s}{\mu+s}} \left(\dfrac{\varepsilon_I}{\omega}\right)^{\frac{\mu}{\mu+s}} & \text{when} \end{cases} \right\} \begin{array}{l} |z| \ll 1 \\[4ex] |z| \gg 1 \end{array} \tag{5.2-30}$$

Clearly with this type of approach the general behavior of a network of impedances may be examined. There is an interesting property associated with the loss angle δ, which is related to the phase difference between the real and imaginary parts of the conductivity via the relation $\tan \delta = \text{Re}(\Sigma)/\text{Im}(\Sigma)$. By considering the scaling laws, we can see that at the percolation threshold the loss angle tends to a universal value δ_c as the frequency tends to zero (see the review article by Clerc et al. 1990):

$$\delta_c = (\pi/2)\, s/(s + \mu) \qquad (p = p_c,\; \omega \rightarrow 0) \; . \tag{5.2-31}$$

Experiments on powder mixtures (spheres of silvered and unsilvered glass) are in agreement with this behavior.

The relation from which the universal value of the loss angle may be found is simple to deduce from the general form of the conductance suggested by Webman et al. (1975), Efros and Skolovskii (1976), and Straley (1976, 1977). It is the continuation in the complex plane of Eq. (5.2-12) and it may be written [11]

$$\Sigma\,(p,\omega) = (R_0)^{-1}\ \left|\,p - p_c\,\right|^{\mu}\ \Phi_{\pm}\!\left(\frac{i\,\omega}{\omega_0}\ \left|\,p - p_c\,\right|^{-(s+\mu)}\right), \qquad (5.2\text{-}32)$$

where $\Phi_{\pm}$ are the scaling functions above and below the percolation threshold.

5.2.4 Response of viscoelastic systems

Percolation has also been commonly used to describe gelation processes. Some details of this idea, which we was mentioned in Sec. 2.8.1, will be given here, as will its application to the mechanics of gels.

The first studies on gelation leading to a statistical description of the sol–gel transition are due to Flory in 1941 (see Flory, 1971). These involved a mean field approach of the "Cayley tree" variety described in Sec. 3.1.1. In 1976, de Gennes and Stauffer suggested replacing this description with a percolation approach (see de Gennes, 1979, and Stauffer et al., 1982).

An example of chemical gelation by *polycondensation*, concerning the formation of aerogels, was given earlier (in Sec. 5.1.4). In this case polymerization occurs through the elimination of a small molecule (water for silica aerogels). In general terms, bi-, tri,- or quadri- functional monomers create one $(f = 2)$, two- or three $(f > 2)$- dimensional lattices. Other processes are equally possible, for example *vulcanization* where bridges form progressively between long chains of polymers, or *additive polymerization* where double or triple bonds are form, allowing the linkage of other clusters.

The earlier dynamic approach can therefore be applied naturally to the mechanical properties of the sol–gel transitions by adopting a percolation model. The rheological properties of a viscoelastic system are in fact characterized by the dynamic viscosity[12] μ and by the complex modulus of elasticity $G = G' + iG''$. The real part $G'(\omega = 0)$ is the modulus of elasticity E of the medium, while its imaginary part G'' representing the dissipation is related to the viscosity by the relation[13] $G'' = \mu\,\omega$, a similar relation to the one linking the imaginary part of Σ to the dielectric constant ε (Sec. 5.2.3). The general expression for G is written the same way as that for Σ :

---

[11] As the variable in the function $\Phi_{\pm}$ is dimensionless, there is a certain freedom in the way the function is written, which explains why there are several (equivalent !) forms for the conductivity Σ.

[12] The viscosity is sometimes denoted η.

[13] This relation is comparable with that of the complex conductivity (Sec. 5.2.3) by making the substitution $\varepsilon/4\pi \Leftrightarrow \mu$.

$$\boxed{G\,(p,\omega) = \; G_0 \; \Phi_{\pm}\left(\frac{i\,\omega}{\Omega_0}\right)} \qquad (5.2\text{-}33)$$

with $\;G_0 \propto \left|p-p_c\right|^{\mu};\quad \Omega_0 = \omega_0 \left|p-p_c\right|^{(s+\mu)}$.

At low frequencies, $\omega \ll \Omega_0$, the beginning of the expansion of Φ may be written

— before formation of the gel, $(p < p_c)$,

$$\Phi_{-}(i\omega/\Omega_0) \cong (i\omega/\Omega_0)\,B_- + (i\omega/\Omega_0)^2\,C_- + \ldots$$

— after the formation of the gel, $(p > p_c)$,

$$\Phi_{+}(i\omega/\Omega_0) \cong A_+ + (i\omega/\Omega_0)\,B_+ + \ldots$$

At high frequencies, $\omega \gg \Omega_0$,

$$\Phi_{\pm}(i\omega/\Omega_0) \propto (i\omega/\Omega_0)^\Delta, \quad \text{where } \Delta = \mu/(s+\mu).$$

These expressions allow the behavior of E and μ at zero frequency to be determined. Finally, as in the previous subsection, the loss angle at threshold tends to a universal value as the frequency tends to zero: $G''/G' \rightarrow \tan(\Delta\pi/2)$. Its dependence on ω is relatively small.

In the case of *chemical gels*, the reaction kinetics are more or less constant and do not display any discontinuity at the gelation threshold. A linear correspondence can be established between the time elapsed from the moment the sol was prepared to its state of gelation (as the kinetics depend on the temperature, the system must be suitably thermalized[14]). Thus, between this situation and the percolation model we have the correspondence $t \Leftrightarrow p$. Gelation occurs at time t_g (gelation time) for a concentration, p_c, of chemical bonds. As t_g is approached, the viscosity increases until it diverges according to a law which (in agreement with what we saw earlier) is conjectured to take the form

$$\mu \propto (t_g - t)^{-s} \qquad (5.2\text{-}34)$$

while the modulus of elasticity E, which was zero up to t_g, becomes positive and obeys a law of the form

$$E \propto (t - t_g)^\mu. \qquad (5.2\text{-}35)$$

This type of behavior has actually been observed by Gauthier-Manuel et al. (1984). The value they obtained for μ was close to 1.9.

In the case of *physical gels*, the bridging density between the chains is fixed by the temperature or by the concentration of ions causing reticulation. The system is in equilibrium and the relevant parameter is the bond concentration between chains. Fairly precise experiments have been carried out

---

[14] For example an epoxy resin (such as *araldite*) hardens in 24 hours at 20°C, 3 to 4 hours at 40°C, and 45 minutes at 80°C.

on some very common biopolymers, the pectins (Axelos and Kolb, 1990), in which the bond concentration is related to the concentration of Ca^{++} ions causing reticulation. This is again a percolation model.

The torque and phase difference produced by the reaction of the gel (or sol) are measured experimentally by a cone maintained in a forced oscillation of small amplitude. The gel is subject to a shearing $\gamma(t)$, the important parameter being the rate of shearing $\dot{\gamma}$, which has the dimension of frequency. Thus, we may expect the behavior of the viscosity as a function of shearing rate to be of the form

$$\mu\,(p,\dot{\gamma}) = \mu_0\,\Psi_{\pm}\left(\frac{i\,\dot{\gamma}}{\Gamma_0}\right)$$

$$\mu_0 \propto \; |p - p_c|^{\mu} \; ; \; \Gamma_0 = \dot{\gamma}_0\,|p - p_c|^{(s + \mu)} \; .$$

(5.2-36)

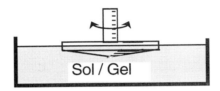

Sol / Gel

Experimental measurements have provided values for pectins, $s \cong 0.82$ and $\mu \cong 1.93$, which agree with a scalar percolation model (i.e., a percolation model together with a model of scalar or entropic vibration). This can be understood in the case of physical gels because here the elasticity of the chains between two bridges is of an entropic nature.

Gels exhibiting tensorial elasticity

Not every gel has exponent values in agreement with percolation models involving scalar elasticity. For instance, we find that for epoxy resins $s \cong 1.4$, for branched polyesters $s > 1.3$ and for gelatine $s = 1.48$, all of which are incompatible with the previous model.

The model may fail for two reasons. First, because the gel's structure cannot be described by a percolation model. For example, hierarchical models have been proposed for aerogels (Maynard, 1989). Second, the elasticity may not be due to entropy. We can see this clearly with aerogels for which the filling fluid is air and where rigidity in bending and compression may be expected. This is just what Vacher et al. (1990) observed (see Sec. 5.1.4 and the phonon–fracton transition). The relation between the spectral dimension d_s and the exponent μ of the modulus of elasticity is unclear, although such a relation does exist in the case of percolation (see Sec. 5.2.2),

$$d_s = \frac{2(\,d\nu - \beta)}{\mu - \beta + 2\nu} = \frac{2\,D}{\mu/\nu + D + 2 - d} \; .$$

(5.2-37)

Moreover, spectral dimensions measured in aerosols concern the region of fractal growth (reaction-limited aggregation) between 10 Å and 100–300 Å; whereas viscoelastic properties (s, μ) are measured in the gel in solution. They necessarily relate to distances greater than size of the fractal clusters (1,000 Å to cm).

For the gelation of silica, Gauthier-Manuel et al.(1987) found $\mu \cong 3$, a value which disagrees with scalar percolation but is closer to the model of Kantor and Webman (1984) for which $\mu = 3.6$, to that of Martin (1986) for which $\mu = 2.67$ or even to the classical model for which $\mu = 3$.

Let us finish by mentioning that in the chemical gels we have considered, percolative structures built out of elementary entities (Fig. 3.5.3), in general by reaction-limited growth, constitute the preliminary stage of gelation. These elementary entities should not strictly be considered as rigid and immobile. The description via a percolation model may thus be improved by adding, as Guyon has suggested, effects due to rotations, deformations, and relaxations of the entities which are themselves fractal and immersed in a viscous (*sol*) liquid.

It would, therefore, still be premature to draw definitive conclusions about gels and their elastic properties.

5.3 Exchanges at interfaces

Many natural or industrial processes take place at an interface between two media. These are particularly important in living systems: blood exchanges oxygen and carbon dioxide with the air surrounding us; and tree roots drain water and mineral salts from the soil by selecting they need. Industrial processes also make great use of interface exchanges. In a battery, charging and discharging cause electrochemical reactions to occur at the interfacial between an electrolyte and an electrode. Similarly, heterogeneous catalytic reactions take place at the catalyst's surface (the reaction products then being on the same side of the interface).

When the reaction at the interface involves a large flow of matter, its efficiency will be higher the larger the surface area of the reaction site. Moreover, to reduce congestion, but above all to increase the rapidity of transport, the volume occupied by the interface should be as small as possible. The problem of maximizing the surface/volume ratio leads in a natural way to fractal structures, or more precisely to porous or tree-like structures (as does the need to optimize the flows of the reacting fluids).

The following problem concerning fractal interfaces then arises: knowing the fractal geometry of the interface, what can be said about the transfer properties across this interface?

Electrodes have recently been the object of a number of studies following the experiments of Le Méhauté and Crépy in 1983 (see Sapoval in *Fractals in*

Disordered Systems, 1991, for a review of this subject, as well as a discussion of various experiments by Rammelt and Reinhard, 1990), but remain little understood. While in a linear response régime, the impedance is commonly observed to behave according to a power law with a nontrivial exponent η at low frequencies:

$$Z \propto (i\omega)^{-\eta} . \qquad (5.3\text{-}1)$$

This is known as *Constant Phase Angle* (CPA) behavior.

Several points concerning transfer across electrodes have been clarified:

— The frequency response of a fractal electrode depends on its electrochemical régime.

— In the diffusion-limited régime the impedance of the electrode depends on its Bouligand–Minkowski dimension (Sec. 1.3.1).

— In the blocking régime the electrode response exhibits constant phase angle (CPA) behavior which in certain special cases has been understood in terms of its geometry. For the moment there is no general description of this phenomenon.

— The microscopic response of a fractal electrode is not proportional to the microscopic transport coefficients, but instead to powers of these parameters. In particular, this can induce errors of interpretation in calculations concerning interface processes.

— Finally, there are exact correspondences between the electrical response of an electrode to a direct current, the diffusive response of a membrane, and the output of a heterogeneous catalyst in a steady state, all having the same geometry.

First of all let us look at the role played by the interfaces in a battery made up of two electrodes separated by an electrolyte.

An electrolytic cell containing a flat electrode of surface area S may be considered as equivalent to a circuit made up of a capacitance $C = \gamma S$, where γ is the specific capacitance of the surface,[14] in parallel with the Faraday resistance R_f. These then act in series with the resistance R_e of the electrolyte. The Faraday resistance is directly proportional to the reciprocal of the rate of electrochemical transfer[15] at the interface and inversely proportional to its surface area ($R_e = r/S$). Two limiting cases may be examined. If there is no electrochemical transfer at the interface, the Faraday resistance is infinite and the electrode is said to be *blocking* or *ideally polarizable*. In the opposite case, it may happen that the *transfer is limited by the diffusion kinetics* of the ions in the electrolyte.

---

[14] This capacitance arises from both the charge accumulation and the surface area. It is obviously defined for a small flat element of unit area.

[15] Like, for instance, in a redox reaction of the type $Fe^{3+} + e^- \rightarrow Fe^{2+}$.

5.3.1 The diffusion-limited regime

In this case, for a flat electrode through which an alternating current of frequency ω flows, the concentration of ions in the neighborhood of the electrode varies locally about a concentration c_0. The variation is governed by a diffusion law, and it can easily be shown (Sapoval et al., 1988) that the region of oscillating concentration extends over a diffusion layer of width $\Lambda_{\mathcal{D}} \cong (\mathcal{D}/\omega)^{1/2}$. The number of ions crossing the surface is proportional to the volume $V(\Lambda_{\mathcal{D}})$ occupied by this diffusion layer. It can then be shown that the electrode has a diffusion admittance of the form $\omega \, c_0 \, V(\Lambda_{\mathcal{D}})$, where c_0 is a capacitance per unit volume or specific diffusive capacitance. c_0 is a parameter which depends on the ion concentration.

Suppose now that the electrode is rough with associated fractal dimension D (Fig. 5.3.1). The size of $V(\Lambda_{\mathcal{D}})$ is then given exactly by the exterior Bouligand–Minkowski dimension of this fractal surface (Sec. 1.3.3):

$$V(\Lambda_{\mathcal{D}}) = S^{D/2} (\Lambda_{\mathcal{D}})^{3-D}, \tag{5.3-2}$$

so that the admittance of the fractal interface may be written

$$\boxed{Y_{\mathcal{D}} \propto c_0 \, S^{D/2} \, \mathcal{D}^{(3-D)/2} \, \omega^{(D-1)/2}} \,. \tag{5.3-3}$$

An admittance of this form displays a CPA structure as follows:

$$Y_{\mathcal{D}} \propto (i\omega)^{\eta}, \quad \text{where } \eta = (D-1)/2. \tag{5.3-4}$$

Thus, in a diffusive regime a fractal electrode exhibits constant phase angle behavior. Of course, the contrary is not true; other sources for this CPA behavior have also been proposed.

Fig. 5.3.1. *External Bouligand–Minkowski dimension associated with an interface. This dimension is defined by* $D = \Delta_{ext}(E) = \lim_{\varepsilon \to 0} [(d - \log V_{\varepsilon}) / \log \varepsilon]$, V_{ε} *being the volume of a layer of width* ε *covering the external surface.*

5.3.2 Response to a blocking electrode

Because blocking electrodes have an infinite Faraday resistance it is impossible for a direct current to flow across them. An electrolytic cell is equivalent to a surface capacitor in series with the resistance of the electrolyte.

There are thus two basic parameters in the system, a specific surface capacitance γ and the resistivity ρ of the electrolyte.

From these two parameters we can construct a natural length for the system:

$$\Lambda(\omega) = (\rho\,\gamma\omega)^{-1} \, . \qquad\qquad (5.3\text{-}5)$$

This length represents the side of a cube of an electrolyte bounded on one side by the electrode and whose volume at a resistance ρ/Λ is equal to the capacitive impedance of the electrode $1/\omega\gamma\Lambda^2$. If we could assume that Λ was the only length involved in this problem [which is not always the case (Meakin and Sapoval, 1991)], the general form of an admittance Y of a fractal regime exhibiting CPA behavior could be written down. Y would then depend only on L/Λ, L being the dimension of the cell, thus

$$Y = f(L/\Lambda) \propto (i\omega)^{\eta} \text{ or consequently } Y \propto (L/\Lambda)^{\eta} \, .$$

Furthermore, a cutoff frequency is reached when $\Lambda > L$. At frequencies lower than this, Y is of the order of L/ρ, so that in the fractal regime we can write

$$\boxed{Y \propto (L/\rho)\,(L/\Lambda)^{\eta} \propto L^{1+\eta}\,\rho^{\eta-1}\,(\gamma\,\omega)^{\eta}} \, . \qquad (5.3\text{-}6)$$

Thus, we have the remarkable fact that the response of such an electrode is not a linear function of the surface area of the cell. This relation agrees with exact results obtained from a Sierpinski electrode (Sapoval et al., 1988). It has also been experimentally verified (Chassaing et al., 1990) for ramified electrodes. In different circumstances, Meakin and Sapoval (1991) have shown that for a whole range of *two-dimensional*, self-similar, ramified electrodes the exponent η is related to the fractal dimension by $\eta = 1/D$, an expression already postulated by Le Méhauté and Crépy in 1983. The impedance Y is then proportional to L^{D_1}, where D_1 ($D_1 = 1$ in $d = 2$) is the information dimension.

Linear response to d_c excitation

To examine the response of a surface with exchange resistance, r, to direct current excitation, invariance properties mean that we need only replace $\gamma\omega$ by r^{-1} in the expression for the admittance, hence

$$Y_c \propto L^{1+h}\,r^{h-1}\,r^{-h} \, . \qquad\qquad (5.3\text{-}7)$$

Membranes and fractal catalysis

It is interesting to note that the properties mentioned above may be applied in a similar way to a wider class of systems governed by laws of the same type. Examples of these are the diffusion of chemicals across membranes, and heterogeneous catalysis on a fractal catalyst. In certain cases one can predict a correspondence between the order and the dimension of the reaction (Sapoval, 1991).

5.4 Reaction kinetics in fractal media

The diffusion-limited chemical kinetics of a bimolecular reaction of type

$$A + B \rightarrow products, \qquad (5.4\text{-}1)$$

between two chemical species A and B in a homogeneous medium, where the As diffuse while the Bs remain motionless, satisfy an evolution equation

$$-\frac{d[A]}{dt} = k\,[A]\,[B]\,, \qquad (5.4\text{-}2)$$

where k is the reaction constant and [A] is the concentration of A. Smoluchowski showed in 1917 that in three dimensions, k is time independent and proportional to the microscopic diffusion constant, $\mathcal{D}$, of species A. More precisely, for a lattice model, k is proportional to the efficiency of a random walker on the lattice, that is to say to the derivative with respect to time of the *mean number of distinct sites* S(t) visited by A and defined in Sec. 5.1.3,

$$k \approx \frac{dS}{dt}\,. \qquad (5.4\text{-}3)$$

Strictly speaking this equation is rigorous only for low concentrations of B (i.e., as [B] $\rightarrow$ 0), but it turns out to be reasonably accurate for realistic concentrations. These equations have been generalized to the case where both chemicals diffuse, and, in particular, to the case of homo-bimolecular reactions,

$$A + A \rightarrow products, \qquad (5.4\text{-}4)$$

whose kinetic equation may be written

$$-\frac{d[A]}{dt} = k\,[A]^2\,. \qquad (5.4\text{-}5)$$

If A is supplied so as to maintain a steady state, the rate R of reagent production can be written

$$R = k\,[A]^2\,, \qquad (5.4\text{-}6)$$

where the coefficient k is time independent. This situation occurs in dimension d = 3 where S(t) = a t (for large enough t, see Sec. 5.1.3).

Heterogeneous kinetics occurring in a fractal structure differ greatly from homogeneous kinetics. Kopelman and his collaborators (Kopelman, 1986) have shown that for the simple reaction

$$A + A \rightarrow products,$$

the *diffusion-limited* reaction rate is

$$k = k_0 \, t^{-h} \, [A]^2, \qquad (t \geq 1)$$

(5.4-7)

where
$$h = 1 - d_s/2 \qquad \text{if} \quad d_s \leq 2,$$
(5.4-8a)
$$h = 0 \qquad \text{if} \quad d_s > 2.$$
(5.4-8b)

These are expressions easily obtained from Eqs. (5.1-22) and (5.1-23), and the preceding remarks (d_s is the spectral dimension).

When the reaction is supplied with A so as to maintain a steady state, the rate R of reagent production can now be written,

$$R = k_0 \, [A]^X .$$

(5.4-9)

The order X of the reaction is no longer equal to two but to

$$X = 1 + \frac{2}{d_s} = \frac{2-h}{1-h} .$$

(5.4-10)

For a percolation cluster structure (d = 3), h ≈ 0.33, and X ≈ 2.5. For a fractal "dust" with spectral dimension $0 < d_s < 1$ the order of the reaction is greater than 3.

Thus in heterogeneous structures we notice a correlation between energy and geometry, which brings about fractal type kinetics.

The results obtained by taking into account scaling invariance are in good agreement with numerical simulations and with experiment. This type of kinetics applies, for example, to excitations fusion experiments in porous membranes, films, and polymeric glasses.

Experiments have been performed on isotopic mixed crystals of naphthalene by Kopelman and his collaborators in the U.S.A., and by Evesque and Duran in France. The mixture of $C_{10}H_{18}$ and $C_{10}D_{18}$ displays a random substitutional disorder and is such that excitations remain localized on the naphthalene clusters at temperatures of 2°K. For triplet excitons:

triplet + triplet → singlet,

the effective percolation threshold in the naphthalene clusters is about 8% mole fraction (corresponding to fourth-nearest-neighbor connectivity in a square lattice). In this experiment the phosphorescence (excited triplet → fundamental singlet transition), which is proportional to the exciton concentration [A], and the fluorescence (excited singlet → fundamental singlet transition), which is proportional to the annihilation rate d[A]/dt, are considered separately. The measured slope is h = 0.35 ± 0.03 in agreement with the results from the percolation model:[16] the disorder here is essentially of geometric origin. On the other hand, experiments carried out at 6°K on triplet exciton fusion in naphthalene inside a porous Vycor glass give a larger value of h, h ≅ 0.44. The

---

[16] In principle the diffusion should be averaged over the infinite cluster (fractal on all scales up to the threshold) and the finite clusters. This provides two contributions of unequal exponent.

structure of Vycor being very different from a percolative structure, but it too involves an energetic disorder superimposed on the geometric one.

Another manifestation of these fractal reaction kinetics is the spontaneous appearance of a segregation (Zeldovich effect) of species A and B in hetero-bimolecular reactions of the type

$$A + B \rightarrow AB\uparrow .$$

This phonemenon is also related to the fact that the future evolution depends on the past and particularly on the initial conditions:

(i) Dependence on the past appears in the rate $k = k_0 \, t^{-h} \, [A]^2$, which shows that, even with equal concentrations of [A], two samples have different reaction rates if the time elapsed since the start ($t = 1$) of the two reactions is different (it is assumed that initially the distributions are random and $k_0 \equiv k_0^r$ is the same). This effect is, of course, related to the fact that the distribution does not remain random and that a time-dependent segregation occurs.

(ii) Order is present in the steady state itself. For example, consider a photochemical reaction over a disordered structure. If the reacting species A is produced by laser impulsion, its initial distribution may be taken to be random: $k_0 \equiv k_0^r$. If we now produce a steady state by permanently illuminating the sample, we find a partial order there (although the supply of A is randomly distributed), the constant k_0 changes, $k_0 \equiv k_0^{ss}$ (at equal initial concentrations), and a simple empirical criterion for the partial order is then given by the ratio $F = k_0^r / k_0^{ss}$.

These heterogeneous reaction kinetics can also be found to occur in chemistry, biology, geology, and solid state physics, as well as in astrophysics and the atmosphere sciences. For instance, the origin of the separation of charges in both colloids and clouds may be attributed to *reactive segregation*.

COLOR PLATE LEGENDS

Page I above: Colony of bacteria, *Bacillus subtillis*, spreading on an agar plate (Matsushita and Fujikawa, 1990). The growth displays a certain similarity to the DLA model, since the spread of the bacteria is governed by the diffusion of the nutritants that they absorb (photograph kindly provided by   Matsushita, Chuo University, Tokyo).

Page I below: Sliding spark forming at the surface of a dielectric (S. Larigaldie in *Fractal Forms*, edited by E. Guyon and H.E. Stanley, Elsevier/North-Holland and Palais de la Découverte, 1991).

Page II above: Electrolytic deposit of copper, obtained by electrolysis in a thin layer of copper sulphate solution situated between two glass plates. As one of these plates has had its surface treated, growth occurs there (photograph kindly provided by Vincent Fleury, Laboratoire PMC, Ecole Polytechnique).

Page II below: Enlargement of the box indicated in the photograph above, demonstrating the self-similar character of the deposit. The fractal dimension was found by V. Fleury to be D = 1.76.

Page III above: Electrolytic deposit of copper showing a similar structure to page II, but using untreated glass plates. Growth occurs throughout the gap. The object shown here has a broadly self-affine structure (photograph kindly provided by Vincent Fleury).

Page III below: Mould of a three-dimensional dissolved structure produced by injecting water under pressure into a cylinder of plaster (through a hole pierced through the center of the cylinder, as can be clearly seen from the mould) (photograph kindly provided by Roland Lenormand and Gérard Daccord, I.F.P., Rueil-Malmaison).

Page IV above: On the left, a photograph of a cauliflower (Broccoli romanesco), displaying an almost deterministic fractal structure. On the right, enlargement by electron microscope showing the smallest scale at which a "pineapple" structure may be observed (photograph kindly provided by François Grey, Risø National Laboratory, Denmark).

Page IV below: Photograph taken from the landsat satellite of Tibetan mountain ranges, showing the horizontal limit of snowfall, and thus displaying a section of a particularly rough natural surface. The image resembles a dendritic growth structure (P. Taponnier, in *Fractal Forms*, edited by E. Guyon and H.E. Stanley, Elsevier/North-Holland and Palais de la Découverte, 1991).

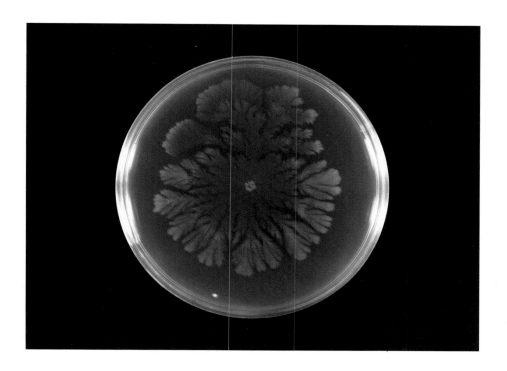

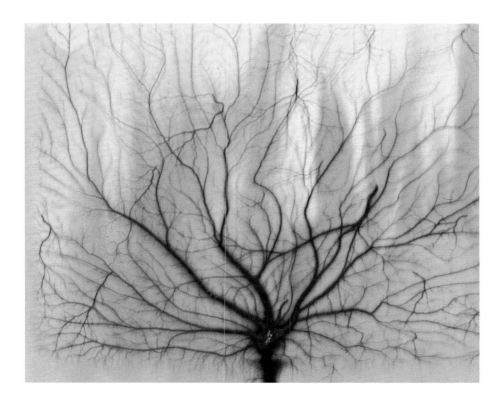

Bibliography

M   General books dealing with mathematical aspects .
PC   General books dealing with physical and chemical aspects.
CR   Proceedings of conferences and schools.
R   Review articles.
f   Simulation films.

PC   ABRAHAM R.H. & SHAW C.D., 1983 — *Dynamics : the Geometry of behavior,* Aerial Press, Santa Cruz.

AHARONY A., 1985 — Anomalous diffusion on percolating clusters. *in:* PYNN & SKJELTORP, p. 289.

AHARONY A., 1987 — Multifractality on percolation clusters. *in:* PYNN & RISTE, p. 163.

ALEXANDER S., 1989 — Vibrations of fractals and scattering of light from aerogels, *Phys. Rev., B* **40**,7953.

ALEXANDER S. & ORBACH R., 1982 — Density of states on fractals: « fractons », *J. Phys. Paris Lett.,* **43**, L625.

ALLAIN C. & JOUHIER B., 1983 — Simulation cinétique du phénomène d'agrégation, *J. Phys. Paris Lett.,* **44**, L421.

PC   AVNIR D., Editeur, 1989 — *The Fractal Approach to the Chemistry of Disordered Systems, Polymers, Colloids and Surfaces*, John Wiley and Sons, New-York.

AXELOS M.A.V. & KOLB M., 1990 — Crosslinked polymers: Experimental evidence for scalar percolation theory, *Phys. Rev. Lett.,* **64**, 1457.

BADII R., 1989 — Conservation laws and thermodynamical formalism for dissipative dynamical systems, *Rivista del Nuovo Cimento,* **12**, 1.

BALIAN R. & SCHAEFFER R., 1989 — Scale-invariant matter distribution in the universe, I. Counts in cells, *Astron. Astroph.,* **220**, 1 ; II. Bifractal behaviour, *Astron. Astroph.,* **226**, 373.

M   BARNSLEY M.F, 1988 — *Fractals everywhere*, Academic Press, Boston.

BARNSLEY M.F, GERONIMO J.S. & HARRINGTON A.N., 1985 — Condensed Julia sets, with an application to a fractal lattice model hamiltonian, *Trans. Amer. Math. Soc.,* **288**, 537

BAUMGÄRTNER A. & HO J.-S., 1990 — Crumpling of fluid vesicles, *Phys. Rev., A* **41**, 5747.

BENGUIGUI L., 1986 — Lattice and continuum percolation transport exponents: experiments in two dimensions, *Phys. Rev., B* **34**, 8176.

BENZI R., PALADIN G., PARISI G. & VULPIANI A., 1984 — On the multifractal nature of fully developed turbulence and chaotic systems, *J. Phys., A* **17**, 3521.

PC   BERGE P., POMEAU Y. & VIDAL Ch., 1988 — *L'Ordre dans le Chaos*, Hermann, Paris.

BERGMAN D., 1989 — Electrical transport properties near a classical conductivity or percolation threshold, *Physica, A* **157**, 72.

BERGMAN D. & IMRY Y., 1977 — Critical behavior of the complex dielectric constant near the percolation threshold of a heterogeneous material, *Phys. Rev. Lett.,* **39**, 1222.

BESICOVITCH A.S., 1935 — On the sum of digits of real numbers represented in the diadic system (On sets of fractional dimensions II.), *Mathematische Annalen,* **110**, 321.

BIRGENEAU R.J. & UEMURA Y.J., 1987 — Spin dynamics in the diluted antiferromagnet $Mn_xZn_{1-x}F_2$, *J. Appl. Phys.,* **61**, 3692.

BLANCHARD A., 1984 — Complex analytic dynamics on the Riemann sphere, *Bull. Amer. Math. Soc. (NS),* **11**, 85.

BLANCHARD A. and ALIMI J.M., 1988 — Practical determination of the spatial correlation function, *Astron. Astrophys.,* **203**, L1.

CR BOCCARA N. & DAOUD M., 1985 — *Physics of Finely Divided Matter,* Springer Verlag, Heidelberg.

BOHR T. & RAND D., 1987 — The entropy function for characteristic exponents, *Physica,* **25**D, 387.

BOUCHAUD E., LAPASSET G. & PLANES J., 1990 — Fractal dimension of fractured surfaces: a universal value?, *Europhys. Lett.,* **13**, 73.

BOULIGAND G., 1929 — Sur la notion d'ordre de mesure d'un ensemble plan, *Bull Sc. Math.,* **II-52**, *185*.

BROADBENTS S.R. & HAMMERSLEY J.M., 1957 — Percolation processes I. Crystals and mazes, *Proc. Cambridge Philos. Soc.,* **53**, 629.

BROLIN H., 1965 — Invariant sets under iteration of rational functions, *Ark. Mat.,* **6**, 103.

BROWN W.D. & BALL R.C., 1985 — Computer simulation of chemically limited aggregation, *J. Phys., A* **18**, L517.

R   BUNDE A., 1986 — Physics of Fractal Structures, *Adv. Solid State Phys.* **26**, 113.

BUNDE A., HARDER H. & HAVLIN S., 1986 — Nonuniversality of diffusion exponents in percolation systems, *Phys. Rev., B* **34**, 3540.

PC BUNDE A. & HAVLIN S., Editeurs 1991 — *Fractals and Disordered Systems,* Springer-Verlag, Berlin.

BUNDE A., ROMAN E., RUSS S., AHARONY A. & BROOKS HARRIS A., 1992 — Vibrational excitations in percolation: localization and multifractality, Phys. Rev. Lett. **69**, 3189.

CAMOIN C., 1985 — Etude expérimentale de suspensions modèles bidimensionnelles, *Thèse de l'université de Provence.*

CHACHATY C., KORB J-P., VAN DER MAAREL J.R.C., BRAS W. & QUINN P., 1991 — Fractal structure of a cross-linked polymer resin: a small-angle x-ray scattering, pulsed field gradient, and paramagnetic relaxation study, *Phys. Rev., B* **44**, 4778.

CHAPUT F., BOILOT J.P., DAUGER A., DEVREUX F. & DE GEYER A., 1990 — Self-similarity of alumino-silicate aerogels, *J. Non-Crys. Sol.,* **116**, 133.

CHARLIER C.V.L., 1922 — How an infinite world may be built up, *Ark. Mat. Astron. Fys.,* **16**, 16.

CLEMENT E., BAUDET C. & HULIN J.P., 1985 — Multiple scale structure of nonwetting fluid invasion fronts in 3D model porous media, *J. Phys. Lett.,* **46**, L1163.

CLEMENT E., BAUDET C., GUYON E. & HULIN J.P., 1987 — Invasion front structure in a 3D model porous medium under a hydrostatic pressure gradient, *J. Phys. D : Appl. Phys.,* **20**, 608.

R   CLERC J.P., GIRAUD G., LAUGIER J.M. & LUCK J.M., 1990 — The a.c. electrical conductivity of binary systems, percolation clusters, fractals, and related models, *Adv. in Phys.,* **39**, 191.

COLEMAN P.H., PIETRONERO L. & SANDERS R.H., 1988 — Absence of any characteristic length in the CfA galaxy catalogue, *Astron. Astrophys.,* **200**, L32.

COLLET P., LEBOWITZ J.L. & PORZIO A., 1987 — The dimension spectrum of some dynamical systems, *J. Stat. Phys.*, **47**, 609.

CONIGLIO A., 1981 — Thermal phase transition of the dilute s-state Potts and n-vector models at the percolation threshold, *Phys. Rev. Lett.*, **46**, 250.

CONIGLIO A., 1982 — Cluster structure near percolation threshold, *J. Phys.*, A **15**, 3829 .

PC COOPER N.G., Editeur 1989 — *From Cardinal to Chaos*, Cambridge University Press.

COURTENS E., PELOUS J., PHALIPPOU J., VACHER R., & WOIGNIER T., 1988 — Brillouin-scattering measurements of phonon-fracton crossover in silica aerogels, *Phys. Rev. Lett.*, **58**, 128.

COURTENS E., VACHER R., PELOUS J. & WOIGNIER T., 1988 — Observation of fractons in silica aerogels, *Europhys. Lett.*, **6**, 245.

R CROQUETTE V., 1982 — Déterminisme et chaos, *Pour la Science*, décembre 1982, p.62.

CURRY J. & YORKE J.A., 1977 — A transition from Hopf bifurcation to chaos: computer experiment with maps in $\mathbb{R}^2$, in *The structure of attractors in dynamical systems*, Springer Notes in Mathematics, **668**, p.48, Springer Verlag.

DACCORD G. & LENORMAND R., 1987 — Fractal patterns from chemical dissolution, *Nature*, **325**, 41-43.

DACCORD G., NITTMAN J. & STANLEY H.E., 1986 — Fractal viscous fingers: Experimental results, *in On Growth and Form*, eds. H.E. STANLEY & N. OSTROWSKY, Martinus Nijhoff, Dordrecht, pp. 203-210).

DAVIDSON D.L., 1989 — Fracture surface roughness as a gauge of fracture toughness: aluminium-particle SiC composites, *J. Mat. Science*, **24**, 681.

DE ARCANGELIS L., REDNER S. & CONIGLIO A., 1985 — Anomalous voltage distribution of random resistor networks and a new model for the backbone at the percolation threshold, *Phys. Rev.*, B **31**, 4725.

DE ARCANGELIS L., REDNER S. & CONIGLIO A., 1986 — Multiscaling approach in random resistor and random superconducting networks, *Phys. Rev.*, B **34**, 4656.

DE ARCANGELIS L. & HERRMANN H.J., 1989 — Scaling and multiscaling laws in random fuse networks, *Phys. Rev.*, B **39**, 2678.

DE ARCANGELIS L. HANSEN A., HERRMANN H.J. & ROUX S., 1989 — Scaling laws in fracture, *Phys. Rev.*, B **40**, 877.

R DE GENNES P.G., 1976 — La percolation : un concept unificateur, *La Recherche*, **72**, 919.

PC DE GENNES P.G., 1979, 1985 — *Scaling Concepts in Polymer Physics*, Cornell Univ. Press, Ithaca, N.Y..

R DE GENNES P.G. & GUYON E., 1978 — Lois générales pour l'injection d'un fluide dans un milieu poreux aléatoire, *J. Mec.*, **17**, 403.

DE GENNES P.G., LAFORE P. & MILLOT J.P., 1959 — Amas accidentels dans les solutions solides désordonnées, *J. Phys. Chem. Solids.*, **11**, 105.

CR DEUTSCHER G., ZALLEN R. & ADLER J., Editeurs 1983 — Percolation structures and Processes, *Ann. Isr. Phys. Soc.*, Vol. 5.

DEUTSCHER G., KAPITULNIK A. & RAPPAPORT M., 1983 — Percolation in metal-insulator systems, *in* Percolation structures and Processes, eds. DEUTSCHER G., ZALLEN R. & ADLER, J., *Ann. Isr. Phys. Soc.* **5**, 207.

DE VAUCOULEURS G., 1970 — The case for a hierarchical cosmology, *Science*, **167**, 1203.

DHAR D., 1977 — Lattices of effectively nonintegral dimensionality, *J. Math. Phys.*, **18**, 577.

DIMOTAKIS P.E., MIAKE-LYE R.C. & PAPANTONIOU D.A., 1983 — Structure and dynamics of round turbulent jets, *Physics of Fluids*, **26**, 3185.

ECKHARDT R., 1989 — extrait de « *Nonlinear Science* » D.K. CAMPBELL, p. 235, *in From Cardinal to Chaos*, ed. N.G. COOPER, Cambridge University Press.

EDEN M., 1961 — , *Proceedings of the 4th Berkeley symposium on mathematical statistics and probabilities*, ed. F. NEYMAN, Université de Californie, Berkeley, **4**, 223.

EFROS A.L. & SHKLOVSKII B.I., 1976 — Critical behaviour of conductivity and dielectric constant near the metal-non-metal transition threshold, *Phys. Stat. Sol. (b)*, **76**, 475.

ELAM W.T., WOLF S.A., SPRAGUE J., GUBSER D.V., VAN VECHTEN D., BARZ G.L. & MEAKIN P., 1985 — Fractal agregates in sputter-deposited NbGe$_2$ films, *Phys. Rev. Lett.*, **54**, 701.

ENGLMAN R. & JAEGER Z., Editeurs 1986 — Fragmentation Form and Flow in Fractured Media, *Ann. Isr. Phys. Soc.*, **8**.

ESSAM J.W., 1972 — Percolation and cluster size, *in Phase transitions and critical phenomena, vol.* **2** p. 197, Academic Press, London.

R   ESSAM J.W., 1980 — Percolation theory, *Rep. Prog. Phys.* **43**, 833.

EVERTSZ C., 1990 — Self-affine nature of dielectric breakdown model clusters in a cylinder, *Phys. Rev., A* **41**, 1830.

M   FALCONER K.J., 1985 — *The Geometry of Fractal Sets*, Cambridge University Press, Cambridge.

M   FALCONER K.J., 1990 — *Fractal Geometry : Mathematical Foundations and Applications*, John Wiley, New York.

FAMILY F., 1990 — Dynamic scaling and phase transitions in interface growth, *Physica, A* **168**, 561.

CR   FAMILY F. & LANDAU D.P., Editeurs 1984 — *Kinetics of Aggregation and Gelation*, North-Holland, Amsterdam.

FATOU P., 1919-1920 — Sur les équations fonctionnelles, *Bull. Soc. Math. France*, **47**, 161 ; **48**, 33 and 208.

PC   FEDER J., 1988 — *Fractals*, Plenum Press, New-York.

FEIGENBAUM M.J., 1978 — Quantitative universality class of nonlinear transformations, *J. Stat. Phys.*, **19**, 25 ; voir aussi, Universal behavior in nonlinear systems, *Los Alamos Science,* Summer 1980.

FLEURY V., ROSSO M., CHAZALVIEL J.N. & SAPOVAL B., 1991 — Experimental aspects of dense morphology in copper electrodeposition, *Phys. Rev., A* **44**, 6693.

PC   FLORY P.J., 1971 — *Principles of polymer chemistry*, Cornell University Press, Ithaca, N.Y. ; les articles originaux ont publiés en 1941 dans *J. Am. Chem. Soc.*, **63**, 3083, 3091, 3096.

FORREST S.R. & WITTEN T.A., 1979 — Long range correlation in smoke aggregates, *J. Phys., A* **12**, L109.

M   FOURNIER D'ALBE E.E., 1907 — *Two new worlds : I. The infra world ; II. The supra world*, Longmans Green, London.

FRISCH U., SULEM P.-L. & NELKIN M., 1979 — A simple dynamical model of intermittent fully developed turbulence, *J. Fluid Mech.*, **87**, 719.

FRISCH U. & PARISI G., 1985 — On the singularity structure of fully developed turbulence. *in Turbulence and Predictability in Geophysical Fluid Dynamics and Climate Dynamics* , eds. M. GHIL, R. BENZI & G. PARISI, North-Holland, New York, pp. 84-88).

GAUTHIER-MANUEL B., GUYON E., ROUX S., GITS S. & LEFAUCHEUX F., 1987 — Critical viscoelastic study of the gelation of silica particles, *J. Phys. (Paris)*, **48**, 869.

GEFEN Y., AHARONY A. & MANDELBROT B.B., 1983 — Phase transitions in fractals. I. Quasi-linear lattices, *J. Phys. A : Math. Gen.*, **16**, 1267.

PC   GLEICK J., 1987 — *Chaos, making a new science,* Cardinal, Sphere books Ltd, Londres.

GOUYET J.F., 1988 — Structure of diffusion fronts in systems of interacting particles, *Solid State Ionics,* **28-30**, 72.

GOUYET J.F., 1990 — Invasion noise during drainage in porous media, *Physica, A* **168**, 581.

GOUYET J.F., ROSSO M. & SAPOVAL B., 1988 — Fractal structure of diffusion and invasion fronts in 3D lattices through gradient percolation approach, *Phys. Rev., B* **37**, 1832.

GRASSBERGER P., 1981 — On the Hausdorff dimension of fractal attractors, *J. Stat. Phys.,* **26**, 173.

R   GUYON E., HULIN J.P. & LENORMAND R., 1984 — Application de la percolation à la physique des milieux poreux, *Annales des Mines,* **191**, n°5 & 6, 17.

HALPERIN B.I., FENG S. & SEN P.N., 1985 — Differences between lattice and continuum percolation transport exponents, *Phys. Rev. Lett.,* **54**, 2391.

HALSEY T.C., JENSEN M.H., KADANOFF L.P., PROCACCIA I. & SHRAIMAN B., 1986 — Fractal measures and their singularities: The characterization of strange sets, *Phys. Rev., A* **33**, 1141.

HAUSDORFF F., 1919 — Dimension und äusseres Mass, *Mathematische Annalen,* **79**, 157.

R   HAVLIN S. & BEN-AVRAHAM D., 1987 — Diffusion in disordered media, *Adv. in Phys.,* **36**, 695-79.

R   HAVLIN S. & BUNDE A., 1989 — Probability densities of random walks in random systems, *Physica, D* **38**, 184.

HAYAKAWA Y., SATO S. & MATSUSHITA M., 1987 — Scaling structure of the growth-probabilty distribution in diffusion-limited aggregation processes, *Phys. Rev., A* **36**, 1963.

HENON M., 1976 — A two-dimensional mapping with a strange attractor, *Communications in Mathematical Physics,* **50**, 69.

HENTSCHEL H.G.E. & PROCACCIA I., 1983 — The infinite number of generalized dimensions of fractals and strange attractors, *Physica,* **8D**, 435.

HENTSCHEL H.G.E. & PROCACCIA I., 1984 — Relative diffusion in turbulent media: The fractal dimension of clouds, *Phys. Rev., A* **29**, 1461.

HERRMANN H.J., 1986 — Geometrical cluster growth models and kinetic gelation, *Phys. Rep.,* **136**, 154-227.

HERRMANN H.J., MANTICA G. & BESSIS D., 1990 — Space-filling bearings, *Phys. Rev. Lett.,* **65**, 3223.

HILFER R. & BLUMEN A., 1986 — On finitely ramified fractals and their extension, *in Fractals in Physics,* eds. L. PIETRONERO & E. TOSSATI, Elsevier Science Pub., p. 33

HONG D;C., STANLEY H.E., CONIGLIO A. & BUNDE A., 1986 — Random-walk approach to the two-component random-conductor mixture: Perturbing away from the perfect random resistor network and random super conductor network limits, *Phys. Rev., B* **33**, 4564

PC   HULIN J.P., CAZABAT A.M., GUYON E. & CARMONA F., Editeurs 1990 — *Hydrodynamics of dispersed media,* Random Materials and Processes, Series Ed. H.E. STANLEY & GUYON E., North-Holland, Amsterdam.

HURST H.E., 1951 — Long-term storage capacity of reservoirs, *Trans. Am. Soc. Civ. Eng.,* **116**, 770.

HURST H.E., BLACK R.P. & SIMAIKA Y.M., 1965 — Long-term storage: an experimental study, *Constable, London.*

ISAACSON J. & LUBENSKY T.C. 1980 — Flory exponents for generalized polymer problems, *J. Phys. Lett.,* **41**, L469.

JENSEN M.H., KADANOFF L.P., LIBCHABER A., PROCACCIA I. & STAVANS J., 1985 — Global universality at the onset of chaos: Results of a forced Rayleigh-Bénard experiment, *Phys. Rev. Lett.,* **55**, 2798.

JONES P. & WOLFF T., 1988 — Hausdorff dimension of harmonic measures in the plane, *Acta Math.,* **161**, 131.

JULIA G., 1918 — Mémoire sur l'itération des fonctions rationnelles, *J. Math. Pures et Appl.,* **4**, 47.

R   JULLIEN R., 1986 — Les phénomènes d'agrégation et les agrégats fractals, *Ann.,*

Télécom., **41** 343.

JULLIEN R. & KOLB M., 1984 — Hierarchical model for chemically limited cluster-cluster aggregation, *J. Phys.,* **17**, L639.

PC JULLIEN R. & BOTET R., 1987 — *Aggregation and Fractal Aggregates,* World Scientific, Singapore.

KANTOR Y., 1989 — Properties of tethered surfaces, *in* Statistical Mechanics of membranes and surfaces, eds. D. NELSON, T. PIRAN & S. WEINBERG, *World Scientific,* Vol. **5**, p. 115.

KANTOR Y. & WEBMAN I., 1984 — Elastic properties of random percolating systems, *Phys. Rev. Lett.,* **52**, 1891.

KAPITULNIK A. & DEUTSCHER G., 1982 — Percolation characteristics in discontinuous thin films of Pb, *Phys. Rev. Lett.,* **49**, 1444.

KARDAR M., PARISI G., & YI-CHENG ZHANG, 1986 — Dynamic scaling of growing interfaces, *Phys. Rev. Lett.,* **56**, 889.

KATZ A.J. & THOMPSON A.H., 1985 — Fractal sandstone pores: Implications for conductivity and pore formation, *Phys. Rev. Lett.,* **54**, 1325.

KERTESZ J. & WOLF D.E., 1988 — Noise reduction in Eden models: II. Surface structure and intrinsic width, *J. Phys., A* **21**, 747.

KESSLER D.A., KOPLIK J. & LEVINE H., 1988 — Pattern selection in fingered growth phenomena, *Adv. in Phys.,* **37**, 255.

KJEMS J. & FRELTOFT T., 1985 — Neutron and X-ray scattering from aggregates, *in:* PYNN & SKJELTORP, p. 133.

KLAFTER J., RUBIN R.J. & SCHLESINGER M.F., Editeurs 1986 — *Transport and Relaxation in Random Material,* World Sci. Press, Singapore.

f KOLB M., 1986 — Aggregation processes, *Film (couleur & son) de 22mn,* Freie Universität Berlin, ZEAM, FU Berlin.

KOLB M. & HERRMANN H.J., 1987 — Surface fractals in irreversible aggregation, *Phys. Rev. Lett.,* **59**, 454.

KOLB M., BOTET J. & JULLIEN R., 1983 — Scaling of kinetically growing clusters, *Phys. Rev. Lett.,* **51**, 1123.

KOLB M. & JULLIEN R., 1984 — Chemically limited versus diffusion limited aggregation, *J. Physique (Paris),* **45**, L977.

KOLMOGOROV A.N., 1941 — The local structure of turbulence in incompressible viscous fluid for very large Reynolds numbers, *C.R. Acad. Sc. URSS,* **31**, 538 (traduction anglaise dans S.K. Friedlander, L. Topper Eds, *Turbulence Classic Papers on Statistical Theory,* Interscience Pub., New York, 1961 ; Dissipation of energy in the locally isotropic turbulence, *C.R. Acad. Sc. URSS,* **32**, 16.

KOLMOGOROV A.N., 1962 — A refinement of previous hypothesis concerning the local structure of turbulence in a viscous incompressible fluid at high Reynolds number, *J. Fluid Mech.,* **13**, 82.

R KOPELMAN R., 1986 — Fractal reaction kinetics, *Science,* **241**, 1620.

KOPELMAN R., 1986 — Rate processes on fractals: theory, simulations, and experiments, *J. Stat. Phys.,* **42**, 185, and references therein.

KRAICHNAN R.H., 1974 — On Kolmogorov's inertial-range theories, *J. Fluid Mech.,* **62**, 305.

LACHIEZE-REY M., 1989 — Statistics of the galaxy distribution, *Int. Journ. Theor. Phys.,* **28**,1125.

CR LAFAIT J. & TANNER D.B., Editeurs 1989 — ETOPIM 2, *Proc. of the 2nd Int. Conf. on Electrical Transport and Optical Properties of Inhomogeneous media,* North-Holland, Amsterdam.

PC LASKAR A.L., BOCQUET J.L., BREBEC G. & MONTY C., Editeurs, 1990 — *Diffusion in materials,* Kluwer Academic Publishers, Dordrecht.

LEE J. & STANLEY H.E., 1988 — Phase transition in the multifractal spectrum of Diffusion-Limited Aggregation, *Phys. Rev. Lett.,* **26**, 2945.

PC   LE MEHAUTE A., 1990 — *Les géométries fractales*, Hermès, Paris.
      LE MEHAUTE A. & CREPY G., 1983 — Introduction to transfer and motion in fractal media: the geometry of kinetics, *Solid State Ionics*, **9&10**, 17.
      LENORMAND R., 1985 — Différents mécanismes de déplacements visqueux et capillaires en milieux poreux : Diagramme de phase, *C.R. Acad. Sci. Paris,* Ser. II, **301**, 247-250.
R   LENORMAND R., 1989 — Application of fractal concepts in petroleum engineering, *Physica, D* **38**, 230.
      LENORMAND R., ZARCONE C. & SARR A., 1983 — Mechanisms of displacement of one fluid by another in a network of capillary ducts, *J. Fluid Mech.*, **135**, 337.
      LENORMAND R.& ZARCONE C., 1985 — Invasion percolation in an etched network: Measurement of a fractal dimension, *Phys. Rev. Lett.*, **54**, 2226.
      LENORMAND R., TOUBOUL E. & ZARCONE C., 1988 — Numerical models and experiments on immiscible displacements in porous media, *J. Fluid Mech.*, **189**, 165.
PC   LESIEUR N., 1987 — Turbulence in fluids, in *Mechanics of fluids and transport processes,* ed. R.J. MOREAU & G.Æ. ORAVAS, Martinus Nijhoff, Dordrecht.
      LEVY P., 1930 — Sur la possibilité d'un univers de masse infinie, *Annales de Physique, p. 184.*
M   LEVY P., 1948, 1965 — *Processus stochastiques et mouvement brownien*, Gauthier-Villars, Paris.
      LEVY Y.E. & SOUILLARD B., 1987 — Superlocalization of electrons and waves in fractal media, *Europhys. Lett.*, **4**, 233.
      LEYVRAZ F., 1984 — Large time behavior of the Smoluchowski equations of coagulation, *Phys. Rev., A* **29**, 854.
      LIBCHABER A., FAUVE S. & LAROCHE C., 1983 — Two parameters study of the routes to chaos, *Physica*, **7**D, 73.
      LOBB C.J. & FORRESTER M.G., 1987 — Measurement of nonuniversal critical behavior in a two-dimensional continuum percolating system, *Phys. Rev., B* **35**, 1899.
      LORENZ E.N., 1963 — Deterministic non-periodic flow, *Journal of Atmospheric Sciences*, **20**, 130.
      LOUIS E., GUINEA F. & FLORES F., 1986 — The fractal nature of fracture, in *Fractals in Physics*, eds. L. PIETRONERO & E. TOSSATI, Elsevier Science Pub., p. 177.
      LOVEJOY S., 1982 — Area-perimeter relation for rain and cloud areas, *Science*, **216**, 185.
PC   MA S., 1976 — *Modern Theory of Critical Phenomena*, Benjamin, New-York.
      MAKAROV N.G., 1985 — On the distorsion of boundary sets under conformal mappings, *Proc. London Math. Soc.*, **51**, 369.
      MANDELBROT B.B., 1967 — How long is the coast of Britain? Statistical self-similarity and fractal dimension, *Science*, **155**, 636.
      MANDELBROT B.B., 1974 — Multiplication aléatoire itérées et distributions invariantes par moyenne pondérée aléatoire, *C.R. Acad. Sc. Paris*, **278**, 289 & 355.
M   MANDELBROT B.B., 1975a — *Les objets fractals : forme, hasard et dimension,* Flammarion, Paris.
      MANDELBROT B.B., 1975b — On the geometry of homogeneous turbulence, with stress on the fractal dimension of the iso-surfaces of scalars, *J. Fluid Mech.*, **72**, 401.
      MANDELBROT B.B., 1976 — Géométrie fractale de la turbulence. Dimension de Hausdorff, dispersion et nature des singularités du mouvement des fluides, *Comptes Rendus* (Paris), **282**A, 119 ; — Intermittent turbulence & fractal dimension: kurtosis and the spectrum exponent 5/3+B, in *Turbulence and Navier Stokes Equations,* ed. R. TEMAN, *Lecture Notes in Mathematics*, **565**, 121.
M   MANDELBROT B.B., 1977 — *Fractals : form, chance and dimension*, W.H. Freeman, San Francisco.
      MANDELBROT B.B., 1980 — Fractal aspects of the iteration $z \rightarrow \lambda z(1-z)$ for complex λ

and z, *Non linear dynamics,* R.H.G. HELLEMAN, *Annals of the New-York Academy of Science*, **357**, 249.

M MANDELBROT B.B., 1982 — *The fractal geometry of nature,* W.H. FREEMAN, New York..

MANDELBROT B.B., 1986 — Self-affine fractal sets, *in: Fractals in Physics,* eds. L. PIETRONERO & E. TOSSATI, North-Holland, Amsterdam, p. 3.

MANDELBROT B.B., 1988 — An introduction to multifractal distribution functions, in *Fluctuations and Pattern Formation,* ed H.E. STANLEY & N. OSTROWSKY, KLUWER, Dordrecht.

M MANDELBROT B.B., 1992 — *Multifractals, 1/f Noise, 1963-1976,* Springer, New York. Plusieurs volumes de compilation sont prévus.

MANDELBROT B.B. & GIVEN J.A., 1984 —Physical properties of a new fractal model of percolation clusters, *Phys. Rev. Lett.,* **52**, 1853.

MANDELBROT B.B., PASSOJA D. E. & PAULLAY A.J., 1984 — Fractal character of fracture surfaces of metals, *Nature,* **308**, 721.

MANDELBROT B.B. & VAN NESS J.W., 1968 — Fractional Brownian motions, fractional noises and applications, *SIAM Rev.,* **10**, 422.

MANDELBROT B.B. & WALLIS J.R., 1968 — Noah, Joseph, and operational hydrology, *Water Resour. Res.,* **5**, 321.

MARTIN J.E., 1986 — Scattering exponents for polydisperse surface and mass fractals, *J. Appl. Cryst.,* **19**, 25.

MATSUSHITA M., SANO M., HAYAKAWA Y., HONJO H. & SAWADA Y., 1984 — Fractal structures of zinc metal leaves grown by electrodeposition, *Phys. Rev. Lett.,* **53**, 286.

MATSUSHITA M., HAYAKAWA Y. & SAWADA Y., 1985 — Fractal structure and cluster statistics of zinc-metal trees deposited on a line electrode, *Phys. Rev. A* **32**, 3814.

MATSUSHITA M. & FUJIKAWA H., 1990 — Diffusion-limited growth in bacterial colony formation, *Physica, A* **168**, 498.

MAYNARD R., 1989 — Elastic and thermal properties of hierarchical structures: Application to silica aerogels, *Physica, A* **157**, 601.

MEAKIN P., 1983 — Formation of fractal clusters and networks by irreversible diffusion-limited aggregation, *Phys. Rev. Lett.,* **51**, 1119.

R MEAKIN P., 1987 — The growth of fractal aggregates, *in:* PYNN & RISTE, p. 45.

MEAKIN P., CONIGLIO A., STANLEY H.E. & WITTEN T.A., 1986 — Scaling properties for the surfaces of fractal and nonfractal objects: An infinite hierarchy of critical exponents, *Phys. Rev., A* **34**, 3325.

MEAKIN P. & SAPOVAL B., 1991 — Random-walk simulation of the response of irregular or fractal interfaces and membranes, *Phys. Rev., A* **43**, 2993.

MENEVEAU C. & SREENIVASAN K.R., 1987a — Simple multifractal cascade model for fully developed turbulence, *Phys. Rev. Lett.,* **59**, 1424.

MENEVEAU C. & SREENIVASAN K.R., 1987b — The multifractal spectrum of the dissipation field in turbulent flow, *in: Physics of Chaos and Systems far from Equilibrium,* eds. MINH-DUONG VAN and B. NICHOLS, North-Holland, Amsterdam.

MENEVEAU C. & SREENIVASAN K.R., 1991 — The multifractal nature of turbulent energy dissipation, *J. Fluid Mech.,* **224**, 429.

METROPOLIS N., STEIN M.L. & STEIN P.R., 1973 — On finite limit sets for transformationon the unit interval, *J. Comb. Theory,* **15**, 25.

MINKOWSKI H., 1901 — Über die Begriffe Länge, Oberfläsche und Volumen, *Jahresbericht der Deutschen Mathematikervereinigung,* **9**, 115.

MURAT M. & AHARONY A., 1986 — Viscous fingering and diffusion-limited aggregates near percolation, *Phys. Rev. Lett.,* **57**, 1875.

NIEMEYER L., PIETRONERO L. & WIESMANN H.J., 1984 — Fractal dimension of dielectric breakdown, *Phys. Rev. Lett.,* **52**, 1033-1036.

NOVIKOV E. & STEWART R.W., 1964 — Turbulence intermittency and fluctuation spectrum of the dissipation energy (in Russian), *Isvestia Akademii Nauk SSR; Seria Geofizicheskaia*, **3**, 408.

OBUKHOV A.M., 1962 — Some specific features of atmospheric turbulence, *J. of Fluid Mech.*, **13**, 77.

R ORBACH R., 1986 — Dynamics of fractal networks, *Science*, **231**, 814.

ORBACH R., 1989 — Fractons dynamics, *Physica, D* **38**, 266.

R PALADIN G. & VULPIANI A., 1987 — Anomalous scaling laws in multifractal objects, *Phys. Rep.,* **156**, 147

PANDE C.S., RICHARDS L.R. & SMITH S., 1987 — Fractal characteristics of fractured surfaces, *J. Mat. Science Lett.,* **6**, 295.

PATERSON L., 1984 — Diffusion-limited aggregation and two-fluid displacements in porous media, *Phys. Rev. Lett.,* **52**, 1621.

PC PECKER J.-C., 1988 — Le ciel est noir, dans L'univers : des faits aux théories, *Pour la Science*, Belin, Paris.

PEEBLES P.J.E., 1989 — The fractal galaxy distribution, *Physica, D* **38**, 273.

M PEITGEN H.O., Editeur 1988 — *The Art of Fractals. A Computer Graphical Introduction*, Springer-Verlag, Berlin.

M PEITGEN H.O.& RICHTER P.J., Editeurs 1986 — *The Beauty of Fractals,* Springer-Verlag, Berlin.

M PEITGEN H.O.& SAUPE D., Editeurs 1988 — *The Science of Fractal Images*, Springer-Verlag, Berlin.

PER BAK, TANG C. & WIESENFELD K., 1988 — Self-organized criticality, *Phys. Rev., A* **38**, 364.

PETERSEN J., ROMAN H.E., BUNDE A. & DIETERICH W., 1989 — Nonuniversality of transport exponents in continuum percolation systems: effects of finite jump distance, *Phys. Rev., B* **39**, 893.

PFEIFER P., AVNIR D. & FARIN D., 1984 — Scaling behavior of surface irregularity in the molecular domain: From adsorbtion studies to fractal catalysts, *J. Stat. Phys.*, **36**, 699.

CR PIETRONERO L. & TOSSATI E., Editeurs 1986 — *Fractals in Physics,* Elsevier, North-Holland, Amsterdam.

CR PIETRONERO L., Editeur 1989 — *Fractals' physical origin and properties*, Plenum Press, New-York.

PIETRONERO L., 1987 — The fractal structure of the universe. Correlation of galaxies and clusters and the average mass density, *Physica, A* **144**, 257.

PO-ZEN WONG, 1988 — The statistical physics of sedimentary rock, *Physics Today,* décembre, 24.

PONTRJAGIN L. & SCHNIRELMAN L., 1932 — Sur une propriété métrique de la dimension, *Ann. of Math.,* **33**, 156.

CR PYNN R. & RISTE T., Editeurs 1987 — *Time-Dependent Effects in Disordered Materials*, Plenum Press, New York.

CR PYNN R. & SKJELTORP A., Editeurs 1985 — *Scaling Phenomena in Disordered Systems*, Plenum Press, New York.

RAMMAL R. & TOULOUSE G., 1983 — Random walks on fractal structures and percolation clusters, *J. Physique Lett.,* **44**, 13.

RAMMELT U. & REINHARD G., 1990 — On the applicability of a constant phase element (CPE) to the estimation of roughness of solid metal electrodes, *Electrochemica Acta,* **35**, 1045.

RAND D.A., 1989 — The singularity spectrum f(α) for cookie-cutters, *Ergod. Theor. & Dynam. Sys.,* **9**, 527.

RICHARDSON L.F., 1961 — The problem of contiguity: an appendix of statistics of deadly quarrels, *General systems yearbook,* **6**, 139.

ROSSO M., GOUYET J.F. & SAPOVAL B., 1985 — Determination of percolation probability from the use of a concentration gradient, *Phys. Rev., B* **32**, 6053.

ROSSO M., GOUYET J.F. & SAPOVAL B., 1986 — Gradient percolation in three dimensions and relation to diffusion fronts, *Phys. Rev. Lett.*, **57**, 3195.

ROUX S. & GUYON E., 1989 — Temporal development of invasion percolation, *J. Phys., A* **22**, 3693.

RUELLE D., 1982 — Repellers for real analytic maps, *Ergod. Theor. & Dynam. Sys.*, **2**, 99.

RUELLE D., 1989 — The thermodynamic formalism for expanding maps, *Commun. Math. Phys.*, **125**, 239.

RUELLE D.& TAKENS F., 1971 — On the nature of turbulence, *Commun. Math. Phys.*, **20**, 167.

RUSS S., ROMAN H. & BUNDE A., 1991 — Vibrational density of states of general two-components mixtures near percolation threshold, *J. Phys. C.*

RYS F.S. & WALDVOGEL A., 1986 — Analysis of the fractal shape of severe convective clouds, *in: Fractals in Physics,* eds. L. PIETRONERO & E. TOSSATI, Elsevier Science Pub., p. 461.

PC SAPOVAL B., 1990 — *Les Fractales,* Edition Diffusion EDITECH, n°**125**.

R   SAPOVAL B., 1991 — Fractal electrodes, fractal membranes and fractal catalysts, *in: Fractals and the Physics of Disordered Systems,* eds. BUNDE A. & HAVLIN S., Springer-Verlag, Berlin.

SAPOVAL B., ROSSO M. & GOUYET J.F., 1985 — The fractal nature of a diffusing front and the relation to percolation, *J. Phys. Lett.,* **46**, L149.

R   SAPOVAL B., ROSSO M. & GOUYET J.F., 1989 — Fractal Physics and Superionic Conductors, *in: Superionic Conductors and Solid Electrolytes: Recent trends,* eds. A. LASKAR & S. CHANDRA, Acad. Press, New-York.

f   SAPOVAL B., ROSSO M., GOUYET J.F. & COLONNA J.F., 1985 — Structure fractale d'un front de diffusion, *film couleur sonore video 12mn,* Imagiciel, 91128 Palaiseau.

SCHAEFER D.W., 1988 — Fractal models and the structure of materials, *MRS Bulletin,* Vol. **XIII**, n°2, 22.

SCHAEFER D.W., WILCOXON J.P., KEEFER K.D., BUNKER B.C., PEARSON R.K., THOMAS I.M. & MILLER D.E., 1987 — Origin of porosity in synthetic materials *in: Physics and chemistry of porous media II,* Ed. J.R. BANAVAR, J. KOPLIK & K.W. WINKLER, Amer. Inst. of Phys., New York.

R   SCHERER G.W., 1990 — Theory of drying, *J. Am. Ceram. Soc.* **73**, 3.

PC SCHUSTER H.G., 1984 — *Deterministic Chaos,* Physik-Verlag, Weinheim.

SEIDEN P.E. & SCHULMAN L.S., 1990 — Percolation model of galactic structure, *Adv. in Phys.,* **39**, 1.

SHAW T.M., 1987 — Drying of an immiscible displacement process with fluid counterflow, *Phys. Rev. Lett.*, **59**, 1671.

SHENDER E.F., 1976 — Thermodynamics of dilute Heisenberg ferroma-gnets near the percolation threshold, *J. Phys. C,* **9**, L309.

SKJELTORP A., 1988 — Fracture experiments on monolayers of microspheres, *in: Random Fluctuations and Pattern Growth : Experiments and Models,* eds. H.E. STANLEY &N. OSTROWSKY, Kluwer Academic Pub. Dordrecht.

M   SMALE S., 1980 — *The mathematics of time : essays on dynamical systems, economic processes and related topics,* Springer Verlag, New York.

SREENIVASAN K.R. & MENEVEAU C., 1986 — The fractal facets of turbulence, *J. Fluid Mech.* **173**, 357.

R   STANLEY H.E., 1985 — Fractal concepts for disordered systems: the interplay of physics and geometry, *in:* PYNN & SKJELTORP, p. 49.

STANLEY H.E. & CONIGLIO A., 1983 — Fractal structure of the incipient infinite cluster in percolation, *in: Percolation structures and Processes.* eds. DEUTSCHER G., ZALLEN R. & ADLER, J., *Ann. Isr. Phys. Soc.,* **5**, 101.

CR STANLEY H.E. & OSTROWSKY N., Editeurs 1985 — *On Growth and Form. Fractal and*

non-fractal patterns in physics, Martinus Nijhoff, Dordrecht.
C R STANLEY H.E. & OSTROWSKY N., Editeurs 1988 — *Random fluctuations and Pattern Growth. Experiments and models*, Kluwer, Dordrecht
C R STANLEY H.E. & OSTROWSKY N., Editeurs 1990 — *Correlations and connectivity : Geometric aspects of physics, chemistry and biology*, Kluwer, Dordrecht.
PC STAUFFER D., 1985 — *Introduction to percolation theory*, Taylor & Francis, London.
STAUFFER D., CONIGLIO A. & ADAM A., 1982 — , *Adv. Polymer Science*, **44**, 103.
STRALEY J.P., 1976 — Critical phenomena in resistor networks, *J. Phys. C*, **9**, 783.
STRALEY J.P., 1977 — Critical exponents for the conductivity of random resistor lattices, *Phys. Rev., B* **15**, 5733.
STRALEY J.P., 1982 — Critical phenomena in resistor networks, *J. Phys. C*, **15**, 2333.
SYKES M.F. & ESSAM J.W., 1964 — Exact critical percolation probabilities for site and bond percolation in two dimensions, *J. Math. Phys.*, **5**, 1117-1127.
PC TAKAYASU H., 1990 — *Fractals in the physical sciences*, Manchester University Press, Manchester and New York.
TRICOT C., 1982 — Two definitions of fractional dimension, *Math. Proc. Camb. Phil. Soc.*, **91**, 57.
TRICOT C., 1988 — Dimension fractale et spectre, *J. Chim. Phys.*, **85**, 379.
VACHER R., WOIGNER T., PELOUS J. & COURTENS E., 1988 — Structure and self-similarity of silica aerogels, *Phys. Rev., B* **37**, 6500.
VACHER R., COURTENS E., CODDENS G., HEIDEMANN A., TSUJIMI Y., PELOUS J. & FORET M., 1990 — Crossovers in the density of states of fractal silica aerogels, *Phys. Rev. Lett.*, **65**, 1008.
VAN DONGEN P.G.J. & ERNST M.H., 1985 — Cluster size distribution in irreversible aggregation at large times, *J. Phys., A* **18**, 2779.
PC VAN DER ZIEL A., 1970 — *Noise, Sources, characterization, measurement*, Prentice-Hall, Inc., Englewood Cliffs, N.J.
VANNIMENUS J., NADAL J.P. & MARTIN H., 1984 — On the spreading dimension of percolation and directed percolation clusters, *J. Phys., A* **17**, L351.
VANNIMENUS J., & NADAL J.P, 1984 — Strip-these for random systems, *Physics Reports*, **103**, 47.
PC VICSEK T., 1989, 1991 — *Fractal growth phenomena*, World Scientific, Singapour.
VON KOCH H., 1904 — Sur une courbe continue sans tangente, obtenue par une construction géométrique élémentaire, *Arkiv för Matematik, Astronomi och Fysik* **1**, 145.
VOLD M.J., 1963 — Computer simulation of floc formation in colloidal suspension, *J. Colloid Sci.*, **18**, 684.
R VOSS R.F., 1985a — Random fractals: Characterization and measurement. *in:* PYNN & SKJELTORP.
R VOSS R.F., 1985b — Random fractal forgeries. *in: Fundamental Algorithms in Computer Graphics*, ed. R.A. EARNSHAW, Springer-Verlag, Berlin (et planches couleur, pp. 13-16).
R VOSS R.F.1988 — Fractals in nature: from characterization to simulation, *in: The Science of Fractal Images*, eds. H.-O. PEITGEN & D. SAUPE, Springer-Verlag, New York Inc.
VOSS R.F., LAIBOVITZ R.B. & ALLESSANDRINI E.I., 1982 — Fractal (scaling) clusters in thin gold films near the percolation threshold, *Phys. Rev. Lett.*, **49**, 1441.
WEBMAN I., JORTNER J. & COHEN M.H., 1985 — Numerical simulation of electrical conductivity in microscopically inhomogeneous materials, *Phys. Rev. B* **11**, 2885.
WEBMAN I. & GREST G.S., 1985 — Dynamical behavior of fractal structures, *Phys. Rev. B* **31**, 1689.
WEITZ D.A. & HUANG J.S., 1984 — Self-similar structures and the kinetics of aggregation of gold colloids, *in: Aggregation Gelation*, eds. F. FAMILY & D.P.

LANDAU, North-Holland, Amsterdam, p. 19.

WEITZ D.A. & OLIVERIA M., 1984 — Fractal structures formed by kinetic aggregation of aqueous gold colloids, *Phys. Rev. Lett.,* **52**, 1433.

WEITZ D.A., LIN M.Y., HUANG J.S., WITTEN T.A., SINSHA S.K., GERTNER J.S. & BALL C., 1985 — Scaling in colloid aggregation, *in:* PYNN & SKJELTORP, p. 171.

WEITZ D.A. & LIN M.Y., 1986 — Dynamic scaling of cluster-mass distribution in kinetic colloid aggregation, *Phys. Rev. Lett.,* **57**, 2037.

WILKINSON D. & WILLEMSEN J.F., 1983 — Invasion percolation: a new form of percolation theory, *J. Phys., A* **16**, 3365.

WILLIAMS T. & BJERKNES R., 1972 — Stochastic model for abnormal clone spread through epithelial basal layer, *Nature,* **236**, 19.

WITTEN T.A. & SANDER L.M., 1981 — Diffusion-limited aggregation, a kinetic critical phenomenon, *Phys. Rev. Lett.,* **47**, 1400.

WITTEN T.A. & SANDER L.M., 1983 — Diffusion-limited aggregation, *Phys. Rev., B* **27**, 5686.

WOIGNER T., PHALIPPOU J., VACHER R., PELOUS J. & COURTENS E., 1990 — Different kinds of fractal structures in silica aerogels, *J. Non-Cryst. Solids,* **121**, 198.

WOLF D.E. & KERTESZ J., 1987 — Surface width exponents for three- and four-dimensional Eden growth, *Europhys. Lett.,* **4**, 651.

R   WOOL R.P., 1988 — *Dynamics and Fractal Structure of Polymer Interfaces,* Internal Symposium on New Trends *in:* Physics and Physical Chemistry, Third Chemical Congress of North America, Toronto, June 5.

WU J., GUYON E., PALEVSKI A., ROUX S. & RUDNICK I., 1987 — Modes de flexion d'une plaque mince au voisinage d'un seuil de percolation, *C.R. Acad. Sci. Paris,* **305**, Série II, 323.

ZALLEN R., 1983 — Introduction to percolation: a model for all seasons, *in:* *Percolation structures and Processes.* eds. DEUTSCHER G., ZALLEN R. & ADLER, J., *Ann. Isr. Phys. Soc.,* **5**, 207.

ZIFF R.M. & SAPOVAL B., 1986 — The efficient determination of the percolation threshold by a frontier generating walk, *J. Phys., A* **19**, 1169.

Index

Aerogels 139, 192
Aerosols 132
Aggregation 130
— in a weak field 142
— macroscopic 140
Aggregation, cluster–cluster 131, 177
— ballistic 180
— diffusion–limited 177
— reaction–limited 179
— reversible 181
Aggregation, particle–cluster 131
— ballistic 131
— diffusion–limited (see DLA)
— reaction–limited (see RLA)
Aggregates 130
Alexander–Orbach conjecture 200
Animals 103
Ant in a labyrinth 197
Anti-DLA 116
Attractor 73, 77
— Lorenz 80
— Hénon 84
— strange 39, 73, 77, 83
— strange, in a Rayleigh–Bénard
   experiment 85
— (multifractality of) 84

Backbone 16, 104, 201
Bacteria 145
Basin of attraction 81
Bifurcations 73, 80
Blocking electrode 212
— response 213

Cantor
— bars 15, 36
— dust 15, 48
— sets 15
Capillary pressure 110
Carpet 16
— Sierpinski 17
Cayley tree 93
Chaos 66, 85
— deterministic 66, 72
Clouds 45, 61
Cluster
— finite 94
— incipient 94
— infinite 94, 146
— percolating 94
— size distribution 102, 179
Coefficient of hydrodynamic dispersion 203
Colloids 132, 134
Compact exploration 191
Conduction 194
— in disordered media 197
Conductivity 195
Connectivity 23

Constant phase angle 212
Contagion 165
Correlation
— function 51, 59, 106
— positive 51
Crossover 167
Curves
— von Koch 11
— Mandelbrot–Given 16
— Peano 13
— f(α) 32, 35, 87

Density of modes (see spectral density)
Deposition 132, 174
— ballistic 176
— electrolytic (see electrodeposition)
— random 174
— random with diffusion 175
Dielectric
— behavior of a composite medium 207
— breakdown 132, 143, 170
— properties 194
Diffusion 118, 188
— anomalous 50
— coefficient, 47, 50, 152, 195
— coefficient (effective), 195
— distinct sites visited by, 191
— in the presence of traps 191
— limited regime 213
Dimension (see fractal dimension)
— Bouligand–Minkowski 6, 9
— box-counting 7, 38, 56
— chemical 24
— covering 4, 5
— critical 98
— critical spectral 191
— generalized (of order q) 33
— global 54
— Hausdorff-Besicovitch 5, 6, 9, 31, 56, 79
— information 35
— latent 54
— local 54
— mass 9
— notion of, 2
— of connectivity 23, 24
— of a path 56
— packing 8
— similarity 22
— spectral 183, 199
— spreading 23, 104
— topological 4, 5, 13, 15
— Tricot 8
Displacement
— of a fluid by another 109
— stable 116
Distance
— chemical 23
— Hausdorff 17

— of connectivity 23
Distribution
— of cluster size 102, 179
— of gains 48
— of galaxies 41
— of voltages 204
DLA 116, 131, 165, 168, 173
Drainage 111, 114
— quasistatic 111
Ductile rupture 62

Eden (see models)
Effective medium 109
Elastic constant 196
Elasticity
— entropic 210
— tensorial 210
Electrodeposition 132, 143, 169
Enstrophy 69
Entropic 186
Epidemics 131
Equation
— Navier-Stokes 67, 80
— Nernst-Einstein 195
— Smoluchowski 180
Exponent(s)
— dynamic percolation 200
— Hölder 28
— Hurst 46
— Lyapounov 81
— of conductivity in disordered media 197,
  199
— percolation 103
— spectral 58

Filtration 143
Fingering
— capillary 116, 118
— viscous 116, 117
Flight
— Brownian 46
— Levy 44
— Rayleigh 46
Fluid(s)
— (displacement by another) 109
— (non)miscibles 109
— (non)wetting 109
Forest fires 131
Fractal catalysis 214
Fractal dimension
— of branching polymers 150
— of Brownian motion 46
— of DLA clusters 167
— of growth processes 131
— of membranes 153, 155
— of percolation clusters 103
— of quadratic map 79
— of RLCA clusters 181
— of self-affine fractals 53
— of two-dimensional diffusion front 123
Fractal music 59
Fractals

— deterministic 10
— heterogeneous 20, 70
— homogeneous 19
— Laplacian 170
— mass 137
— multiscalar 14
— nonuniform 18, 35
— random 19
— pore 137
— self-affine 53
— surface 137, 140
Fractons 183, 187, 192
Fractures 45, 62
Fragile rupture 62, 65
Front
— diffusion 118
— invasion 118
— with interaction 125
Function
— Brownian (of a point) 49
— correlation 51, 59, 106
— fractional Brownian 49, 57
Functionnality 146

Galaxies 41
Gasket 16
— Sierpinski 17, 184, 196
Gel(s) 146, 151
— chemical 151, 209
— physical 151, 209
— with tensorial elasticity 210
Gelation time 151
Generator 11, 15, 19,
Growth (see models)
— bacterial 146
— of percolation cluster 164

Imbibition 111, 114
Imprecision 23
Increments
— correlated 51
— independent 51
Initiator 11
Injection 131, 132
— in a porous medium 112, 145, 170
— in the presense of gravity 129
Interfaces
— exchanges at, 211
Intermittence 73
Internal similarity 21
Invariance
— dilation 21
— scale 21
— under internal similarity 21

Koch island 12
Kolmogorov scale 67, 68

Lacunarity 23, 25
Law(s)
— Darcy 109
— dynamical scaling 160, 178

— Poiseuille 110
— Porod 139
— scaling 12, 102, 149, 159, 178
Layers
— deposited by sputtering 141
— evaporated 105
Legendre transform 35
Length
—connectivity 99
—correlation 99, 153
Loss angle 207, 208

Map
— first return 74, 75
— iterated 81
— quadratic 75, 78
Matching pair 124
Mean field 96, 109, 147, 148, 150, 155
Measure
— α-dimensional Hausdorff measure 2, 6, 8, 35, 36
— binomial fractal 27, 71, 72
— harmonic 172
— multifractal 26, 86
— multifractal, in a random resistor network 203
— multifractal, on a set of points 38
— multinomial fractal 30
— support of 26, 36
Media
— disordered 89, 197
— porous 107
Membranes 146, 152, 214
— fluid 153
— solid 154
Menger sponge 18
Method
— box-counting 7, 53, 62
— disjointed balls 8
— dividers' 8
— R/S analysis 46
Minkowski sausage 6
Modes of vibration 183, 191
Model(s)
— "absolute curdling" 69
— ß 69
— Coniglio-Stanley 201
— deposition 174
— DLA 165
— Eden 132, 157
— growth 157
— hierarchical 43, 204
— Kolmogorov 69
— log-normal 72
— Lorenz 80
— off-lattice 166
— random ß 69
— turbulence 20
— Williams et Bjerknes 163
— Witten and Sander 165
Modulus of elasticity 209

Monophasic flow 108
Motion
— antipersistant 51
— Brownian 45, 52, 56
— Brownian scalar 48
— fractional Brownian 45, 49, 52
— persistant 51
Multifractal 19, 26, 70, 203
Multifractality 26, 84, 172
Multiscale fractals 14

Natural metric 24
Network
— Bethe 93
— random resistor 92, 197, 201
— random superconductor 198
Noise
— during fluid injection 130
— fractionnal Gaussian 50
— Gaussian 59
— in $1/f^{\alpha}$ 124
— white 50, 52
Noise reduction 160
Number
— capillary 116
— Prandtl 81
— Rayleigh 79
— Reynolds 67, 68

Olbers' paradox 43

Pendulum, simple forced 73
Percolation 89, 211
— bond 91
— stirred 141
— continuous 205
— gradient 119
— invasion 113, 115, 117
— over regular networks 92
— scalar 211
— site 91
— threshold 91, 93, 120
Periodically struck rotator 74
Permeability 107
Phase portrait 74
Phonons-fractons 183, 192
— (transition) 193
Poincaré section 75
Polycondensation 208
Polymerization (additive) 208
Polymers 146
— branching 146, 150
— in a bad solvent 149
— in a good solvent 148
— linear 146
— molten 150
— without solvent 149
Porosity 107
— critical 107
Porous media (ideal) 125
Prefractal 12

Probability of survival 192

Radius of gyration 99, 153, 155
Ramification 23, 25, 105
Random walks 188
— self-avoiding 147
Rayleigh-Bénard instability 79
Reaction kinetics in fractal media 215
Reactive segregation 217
Regime
— diffusion-limited aggregation (see
  aggregation)
— diffusion-limited exchange 213
— quasiperiodic 85
Relation
— dispersion 187, 191
— mass-radius 13
— perimeter-area 61, 65
Reliefs (mountainous) 45, 57
Renormalization 184, 196
Renormalization group 184
Response of a blocking electrode 213
Return to the origin 189
RLA 131
RLCA 136
Rough surfaces 174

Scaling law correction 14, 23
Scattered universe 43
Scattering at small angles 137, 148
Sedimentation 142
Self-affine 8, 49, 52-6, 172, 174
Self-affinity 52, 161
Self-organised criticality 130
Self-similar 22, 54, 55
Sensitivity to initial conditions 72, 83
Set
— Cantor 15
— Julia 81, 82
— Mandelbrot 81, 82, 83
— two-scale Cantor 36
— zero 53, 58
Sol 135, 146, 209

Spectral density 58, 183
— of fractional Brownian motion 58
State density 183
Subdiffusion 50
Subharmonic cascade 76, 79
Superdiffusion 50
Systems
— conservative 73
— dissipative 73
— Hamiltonian 73

Termite 198
Thermoconvection 80
Thickening 6
Tortuosity 24
Transfer (diffusion-limited) 212
Transition
— crumpling 153
— metal-insulator 197
— metal-superconductor 197
— phonons-fractons 193
— uncoupling 154
— viscoelastic 194
Transport 194
— continuous percolation 205
Tumors 131
Turbulence 66
— developed 66
— Kolmogorov model 69
— strong 66
— two-dimensional 67
— weak 66
Θ point 149

Universality 94, 106, 107

Vibrations (scalar) 186
Viscoelastic systems (response of) 208
Viscosity (dynamic) 67, 208
Visited sites 191, 215
Vorticity 66
Vulcanization 149, 208

MASSON, Éditeur,
120, Boulevard Saint-Germain
75280 Paris Cedex 06
Dépôt légal : mai 1996

Imprimerie Nouvelle
45800 Saint-Jean-de-Braye
N° d'imprimeur : 30019
Dépôt légal : avril 1996